STUDY GUIDE BOOK

Media Makeup

by HEE-KYUNG, LIM. HYUN-JEONG, SEOK
Professor of the Department of Beauty Art, Seokyung University

HUO TAO
School of Film, Television & Communication, Xiamen University of Technology

SEO KYEONG UNIVERSITY SINCE 1947 서경대학교 글로벌 미용교재개발 연구회

청구문화사

Contents

Lesson 1

미디어 분장개요

OverView of Media Makeup

Lesson 1. 미디어 분장개요 OverView of Media Makeup

1. 미디어 분장 이론 Theory of Media Makeup

■ 미디어 분장이란? What is Media Makeup?

- 영화, 텔레비전, 연극, 사진 촬영 등 다양한 미디어 매체에 사용되는 분장을 의미한다.
- 배우가 연기하는 캐릭터 인물의 외형을 상황에 맞게 리얼리티를 살려 하는 분장을 의미한다.
- 다양한 미디어 환경에서 아나운서, 배우, 모델, 가수 등의 외모를 효과적으로 연출할 수 있다.

■ 미디어 분장의 역할 The role of Media Makeup

- 배우의 역기를 보조하여 감정 및 캐릭터 표현을 더욱 강하게 어필하는 역할을 한다.
- 환상과 리얼리즘 결합으로 현실적이지 않은 캐릭터나 설정을 리얼리티를 살려 현실처럼 보이게 만드는 역할을 한다.
- 현실적인 분장 표현으로 영화, 드라마에서 사건의 진정성을 높여 관객에게 강렬한 몰입감을 제공한다.

1) 텔레비전 Television

- 텔레비전은 영상과 소리를 송출하여 다양한 콘텐츠를 시청할 수 있는 장치이다.

□ TV 드라마분장
TV Drama Makeup

TV 드라마 분장 이미지 삽입
Insert TV Drama Makeup Image.

□ TV 사극분장
TV Historical Drama Makeup

TV 사극 분장 이미지 삽입
Insert Historical TV Drama Makeup Image.

□ 예능프로그램 진행자 분장
Variety Show Host Makeup

예능프로그램 진행자 분장 이미지 삽입
Insert TV Show Host Makeup Image.

□ TV 앵커 분장
TV Anchor Makeup

TV 앵커 분장 이미지 삽입
Insert TV Anchor Makeup Image.

☐ 음악프로그램 진행자 분장
Music Program Host Makeup

음악프로그램 진행자 분장 이미지 삽입 Insert Music Program Host Makeup Image.

☐ 어린이프로그램 진행자 분장
Children's Program Host Makeup

어린이프로그램 진행자 분장 이미지 삽입 Insert Children's Program Host Makeup Image.

2) 영화 Film

- 움직이는 이미지를 통해 이야기를 전달하는 예술 및 대중 매체로 관객에게 시각적, 청각적 경험을 제공한다.
- 오락적 기능뿐만 아니라 사회적, 문화적 메시지를 전달하는 중요한 매체이다.
- 연기, 대사, 음악, 편집, 촬영 기법, 시각 효과 등이 결합된 복합적인 예술 형식이다.

☐ 드라마 장르 영화 분장
Drama Genre Movie Makeup

드라마 장르 영화 분장 이미지 삽입 Insert Drama Genre Movie Makeup Image.

☐ 코미디 장르 영화 분장
Comedy Genre Movie Makeup

코미디 장르 영화 분장 이미지 삽입 Insert Comedy Genre Movie Makeup Image.

☐ 로맨스 장르 영화 분장
Romance Genre Movie Makeup

로맨스 장르 영화 분장 이미지 삽입 Insert Romance Genre Movie Makeup Image.

☐ 뮤지컬 장르 영화 분장
Musical Genre Movie Makeup

뮤지컬 장르 영화 분장 이미지 삽입 Insert Musical Genre Movie Makeup Image.

□ 역사 장르 영화 분장
Historical Genre Movie Makeup

역사 장르 영화 분장 이미지 삽입
Insert Historical Genre Movie Makeup Image.

□ 서부극 장르 영화 분장
Western Genre Movie Makeup

서부극 장르 영화 분장 이미지 삽입
Insert Western Genre Movie Makeup Image.

□ 전쟁 장르 영화 분장
War Genre Movie Makeup

전쟁 장르 영화 분장 이미지 삽입
Insert War Genre Movie Makeup Image.

□ 공포 장르 영화 분장
Horror Genre Movie Makeup

공포 장르 영화 분장 이미지 삽입
Insert Horror Genre Movie Makeup Image.

□ 좀비 장르 영화 분장
Zombie Genre Movie Makeup

좀비 장르 영화 분장 이미지 삽입
Insert Zombie Genre Movie Makeup Image.

□ 스릴러 장르 영화 분장
Thriller Genre Movie Makeup

스릴러 장르 영화 분장 이미지 삽입
Insert Thriller Genre Movie Makeup Image.

☐ **액션 장르 영화 분장**
Action Genre Movie Makeup

범죄 장르 영화 분장 이미지 삽입
Insert Crime Genre Movie Makeup Image.

☐ **범죄 장르 영화 분장**
Crime Genre Movie Makeup

액션 장르 영화 분장 이미지 삽입
Insert Action Genre Movie Makeup Image.

☐ **모험 장르 영화 분장**
Adventure Genre Movie Makeup

모험 장르 영화 분장 이미지 삽입
Insert Adventure Genre Movie Makeup Image.

☐ **판타지 장르 영화 분장**
Fantasy Genre Movie Makeup

판타지 장르 영화 분장 이미지 삽입
Insert Fantasy Genre Movie Makeup Image.

☐ **공상 과학 영화 장르 영화 분장**
Science fiction Genre Movie Makeup

공상 과학 영화 장르 영화 분장 이미지 삽입
Insert Science Fiction Genre Movie Makeup Image.

☐ **다큐멘터리 장르 영화 분장**
Documentary Genre Movie Makeup

다큐멘터리 장르 영화 분장 이미지 삽입
Insert Documentary Genre Movie Makeup Image.

2. 시나리오 및 캐릭터 분석 Scenario and Character Analysis

■ 시나리오 분석

1) 시나리오 분석 Scenario analysis

- 영화, 연극, 드라마 등의 시나리오(대본)를 깊이 있게 해석하고 평가하는 과정이다.
- 스토리와 캐릭터를 이해하고, 작품의 메시지를 효과적으로 전달하기 위해 중요한 과정이다.

□ 줄거리 분석 Plot Analysis	
기(起) Introduction	
승(承) Development	
전(轉) Turn	
결(結) Conclusion	

□ 시대 및 배경 분석 Time Period and Setting Analysis	
시대 분석 Analysis of the Times.	
배경분석 Background Analysis.	

2) 캐릭터 분석 Character Analysis

- 이야기 속 등장인물의 성격, 동기, 그리고 행동을 파악하는 과정이다.
- 이야기 전체에서 맡는 각각의 역할을 깊이 있게 이해하고 해석하는 과정이다.
- 영화, 연극, 드라마 등 모든 이야기에서 매우 중요한 작업이다.
- 캐릭터가 스토리에서 어떤 역할을 하는지, 어떻게 발전하는지, 캐릭터 간의 내적, 외적 갈등이 무엇인지 알 수 있다.

□ 등장인물 분석 Character Analysis		
▸ 기본정보 Basic Information	이 름 Name	
	나 이 Age	
	직 업 Occupation	
	사회적 위치 Social Status	
▸ 역 할 Role	주 연 Main Role	
	조 연 Supporting Role	
	선한 역할 Heroic Role	
	악한 역할 Villainous Role	
▸ 성 격 Personality	성격적 특성 Personality Traits	
	장 점 Strengths	
	단 점 Weaknesses	
	동 기 Motivation	
▸ 외적 표현 Physical Appearance	얼굴 특징 Facial Features	
	외 모 Looks	
	건강 상태 Health Condition	
	행동, 습관 Actions and Habits	
▸ 캐릭터의 관계 Relationships	주요관계 Key Relationships	
	영향력 Influence	
▸ 캐릭터의 발전 Character Development	변화와 성장 Change and Growth	
	긍정적인 Positive Development	
	부정적인 Negative Development	
	동 기 Motivation	
▸ 배경 Backstory	과거의 경험 Past Experiences	
	문화적 배경 Cultural Background	
	사회적 배경 Social Background	

3) 캐릭터 적용 분장 Character Makeup Application

- 기본 분장 Basic Makeup
- 상처 분장 Scar Makeup
- 특수 효과 분장 Special Effect Makeup
- 보철 분장 Prosthetic Makeup

<table>
<tr><th>캐릭터 명
Name of the character</th><th>내용
Content</th><th>설명
Description</th></tr>
<tr><td rowspan="3">배역
Role

배우명
Actress Name</td><td>상황 분석
Situation Analysis</td><td></td></tr>
<tr><td>적용 분장
Applied Makeup</td><td></td></tr>
<tr><td>분장 의도
Makeup Intention</td><td></td></tr>
</table>

4) 미디어의 제작 환경 Production Environment of Media

- 미디어 제작을 위한 촬영 현장.

실내 세트장 Indoor Shooting Sets	오픈 스튜디오 Open Shooting Studios	로케이션 On-location
실내 세트장 이미지 삽입 Insert Indoor Shooting Set Image.	오픈 스튜디오 이미지 삽입 Insert Open Studio Image.	로케이션 이미지 삽입 Insert On-location Image.

Lesson 2

조명과 분장재료
Lighting & Makeup Materials

Lesson 2. 조명과 분장 재료 Lighting & Makeup Materials

1. 색 Color

- 빛이 물체에 반사되거나 투과될 때 인간의 눈에 보이는 시각적 속성을 말한다.

■ 색 이미지 Color Image

- 색 이미지란 특정 색을 통해 전달되는 시각적 인상이나 분위이다.
- 이미지에서 색이 주는 감정적, 상징적 의미를 포함한다.
- 디자인, 예술, 마케팅 등에서 중요한 역할을 한다.
- 색상 조합이나 톤에 따라 특정한 느낌을 전달할 수 있다.

□ 난색 Warm Colors

- 따뜻하고 활기찬 느낌을 주는 색상이다.
- 열정적이고 에너지 넘치는 분위기 형성한다.
- 시각적으로 가까워 보이는 효과를 준다.

□ 주요색상 Main Color

-
-

□ 중간색 Neutral Colors

- 난색과 한색의 중간에 위치한 색상이다.
- 따뜻함과 차가움이 혼합된 느낌을 준다.
- 난색과 한색 성질의 균형을 잡아 다양한 연출이 가능하다.

□ 주요색상 Main Color

-
-

□ 한색 Cool Colors

- 차갑고 조용한 느낌을 주는 색상이다.
- 시각적으로 멀리 있는 것처럼 보인다.
- 차분하고 안정적인 분위기를 조성한다.

□ 주요색상 Main Color

-
-

---

2) 명도 Lightness/Brightness

- 색의 밝고 어두운지의 정도를 나타낸다.

명도 단계 이미지 삽입
Insert Brightness levels Image.

3) 채도 Saturation

- 선명하고 탁한 정도를 나타낸다.

채도 단계 이미지 삽입
Insert a Saturation levels Image.

4) 빛의 3원색

The Three Primary Colors of Light

-
-
-

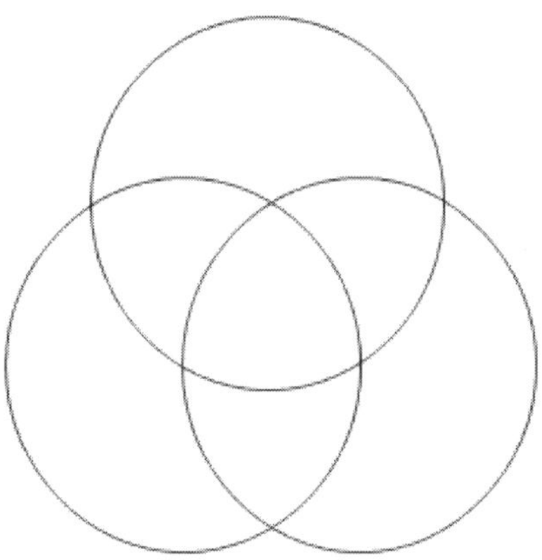

5) 색의 3원색

The Three Primary Colors of Pigment

-
-
-

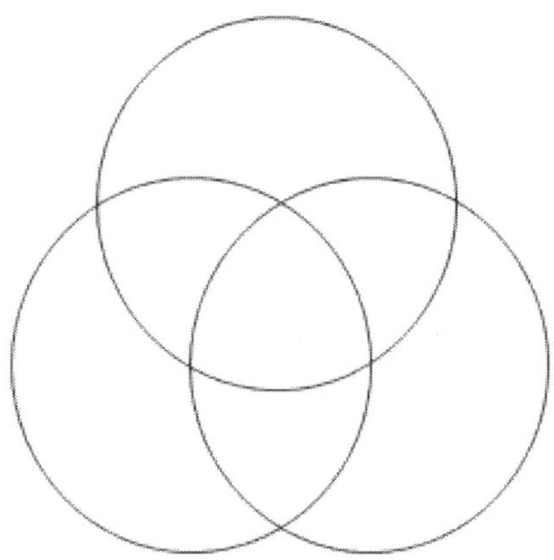

2. 조명 Lighting

■ 조명

- 빛을 사용하여 공간이나 물체를 밝히는 것이다.

1) 조명의 종류 Types of Lighting

주광 Key Light

- 피사체를 주로 비추는 메인 조명

주광 조명 이미지 삽입
Insert Key Light Image.

역광 Back Light or Rim Light

- 피사체 뒤에서 비추는 조명

역광 조명 이미지 삽입
Insert Backlight Image.

보조광 Fill Light

- 키라이트로 인해 생기는 강한 그림자를 완화시키기 위한 조명

보조광 조명 이미지 삽입
Insert Fill Light Image.

하이키 조명 High Key Lighting

- 그림자가 거의 없고, 전체적으로 부드럽고 밝은 느낌을 주는 조명

하이키 조명 이미지 삽입
Insert High Key Lighting Image.

탑라이트 Top Light

- 피사체 위쪽에서 비추는 조명

탑라이트 조명 이미지 삽입
Insert Top Light Image.

엣지라이트 Edge Light

- 피사체의 가장자리를 밝히는 조명

엣지라이트 조명 이미지 삽입
Insert Edge Light Image.

로우키 조명 Low Key Lighting

- 어두운 그림자와 강한 명암 대비의 조명

실루엣 조명 Silhouette Lighting

- 배경에만 빛을 비춰 피사체가 어둡게 보이도록 만드는 조명

로우키 조명 이미지 삽입 Insert Low Key Lighting Image.	실루엣 조명 이미지 삽입 Insert Silhouette Lighting Image.

자연광 Natural Lighting

- 햇빛과 같은 자연스러운 빛을 이용한 조명

자연광 조명 이미지 삽입1 Insert Natural Light Image1.	자연광 조명 이미지 삽입2 Insert Natural Light Image2

컬러 조명 Color Lighting

- 특정 색깔의 조명

빨간색 조명 이미지 삽입 Insert Red Lighting Image.	파란색 조명 이미지 삽입 Insert Blue Lighting Image.	노란색 조명 이미지 삽입 Insert Yellow Lighting Image.
녹색 조명 이미지 삽입 Insert Green Lighting Image.	보라색 조명 이미지 삽입 Insert Purple Lighting Image.	분홍색 조명 이미지 삽입 Insert Pink Lighting Image.

2) 조명의 색온도 Color Temperature of Lighting

- 조명색의 색온도에 따라 따뜻하거나 차가운 분위기를 연출할 수 있다.

■ 색온도의 범위 Range of Color Temperature

□ 낮은 색온도
Low Color Temperature

- 따뜻한 빛
- 약 1,000K ~ 3,500K
- 노란빛 또는 붉은빛을 띤다.
- 따뜻하고 아늑한 느낌

□ 중간 색온도
Medium Color Temperature

- 중립적인 빛
- 약 3,500K ~ 5,000K
- 자연광과 비슷한 중립적인 빛

□ 높은 색온도
High Color temperature

- 차가운 빛
- 약 5,000K ~ 10,000K
- 파란빛 또는 하얀빛을 띤다.
- 차가운 느낌과 선명한 색상 제공

3) 분장 색상과 조명 Makeup Colors and Lighting

메이크업 색상 Makeup Colors	✍ 빨간색 Red Color	✍ 노란색 Yellow Color	✍ 녹색 Green Color	✍ 파란색 Blue Color	✍ 보라색 Violet Color
☀ 빨간색조명 Red Lighting	흐려짐 Fades	레드 Red	어두워짐 Darkens	옅은 레드 Light Red	옅은 레드 Light Red
☀ 오렌지 조명 Orange Lighting	밝아짐 Brightens	조금 흐려짐 Slightly Fades	어두워짐 Darkens	어두워짐 Darkens	밝아짐 Brightens
☀ 옐로 조명 Yellow Lighting	화이트 White	화이트 또는 흐려짐 White or Fades	어두워짐 Darkens	바이올렛 Violet	핑크 Pink
☀ 그린 조명 Green Lighting	어두워짐 Darkens	어두운 그레이 Dark Gray	옅은 그린 Light Green	밝아짐 Brightens	옅은 블루 Light Blue
☀ 블루 조명 Blue Lighting	어두운 그레이 Dark Gray	어두운 그레이 Dark Gray	어두운 그린 Dark Green	옅은 블루 Light Blue	어두워짐 Darkens
☀ 바이올렛 조명 Violet Lighting	블랙 Black	어두운 그레이 Dark Gray	어두운 그레이 Dark Gray	보라색 Violet	매우 옅어짐 Very Faint

출처: 교육부(2019). 미디어 메이크업(LM12010348_18v4). 한국직업능력개발원. p.28. 인용

3. 미디어 분장 재료 Media Makeup Materials

1) 도구 Tools

□

•
•
•

제품이미지 삽입
Insert Product Image.

□

•
•
•

제품이미지 삽입
Insert Product Image.

□

•
•
•

제품이미지 삽입
Insert Product Image.

□

•
•
•

제품이미지 삽입
Insert Product Image.

2) 기초 화장품 Skincare Products

□ 클렌징
Cleansing

•
•
•

제품이미지 삽입
Insert Product Image.

□ 토너
Toner

•
•
•

제품이미지 삽입
Insert Product Image.

□ 에센스/세럼
Essence/Serum

•
•
•

제품이미지 삽입
Insert Product Image.

□ 모이스처라이저
Moisturizer

•
•
•

제품이미지 삽입
Insert Product Image.

□ 아이크림
Eye Cream

•
•
•

제품이미지 삽입
Insert Product Image.

□ 선크림
Sun Cream

•
•
•

제품이미지 삽입
Insert Product Image.

3) 색조 화장품 Makeup Products

☐ 메이크업 베이스
Makeup Base

•
•
•

제품이미지 삽입
Insert Product Image.

☐ 파운데이션
Foundation

•
•
•

제품이미지 삽입
Insert Product Image.

☐ 파우더
Powder

•
•
•

제품이미지 삽입
Insert Product Image.

☐ 컨실러
Concealer

•
•
•

제품이미지 삽입
Insert Product Image.

☐ 쉐딩/하이라이터
Shading/Highlighter

•
•
•

제품이미지 삽입
Insert Product Image.

☐ 블러셔
Blush

•
•
•

제품이미지 삽입
Insert Product Image.

☐ 아이라이너
Eyeliner

•
•
•

제품이미지 삽입
Insert Product Image.

☐ 아이섀도
Eyeshadow

•
•
•

제품이미지 삽입
Insert Product Image.

☐ 마스카라
Mascara

•
•
•

제품이미지 삽입
Insert Product Image.

☐ 립스틱/립글로스
Lipstick/Lip Gloss

•
•
•

제품이미지 삽입
Insert Product Image.

4) 특수 효과 분장 제품 Special Effects Makeup Products

■ 알코올베이스 제품
Alcohol-Based Products

☐

•
•
•

제품이미지 삽입
Insert Product Image.

☐

•
•
•

제품이미지 삽입
Insert Product Image.

☐

•
•
•

제품이미지 삽입
Insert Product Image.

☐

•
•
•

제품이미지 삽입
Insert Product Image.

☐

•
•
•

제품이미지 삽입
Insert Product Image.

■ 수성베이스 제품
Water-Based Products

☐

•
•
•

제품이미지 삽입
Insert Product Image.

☐

•
•
•

제품이미지 삽입
Insert Product Image.

☐

•
•
•

제품이미지 삽입
Insert Product Image.

☐

•
•
•

제품이미지 삽입
Insert Product Image.

☐

•
•
•

제품이미지 삽입
Insert Product Image.

Lesson 3

스트레이트 분장
Straight Makeup

Lesson 3. 스트레이트 분장 Straight Makeup

1. 스트레이트 메이크업 Straight Makeup

-
-
-
-

2. 자료조사 Research

스트레이트 메이크업 이미지 삽입
Insert Straight Makeup Image.

3. 분장 디자인 - 정면 Makeup Design - Front View

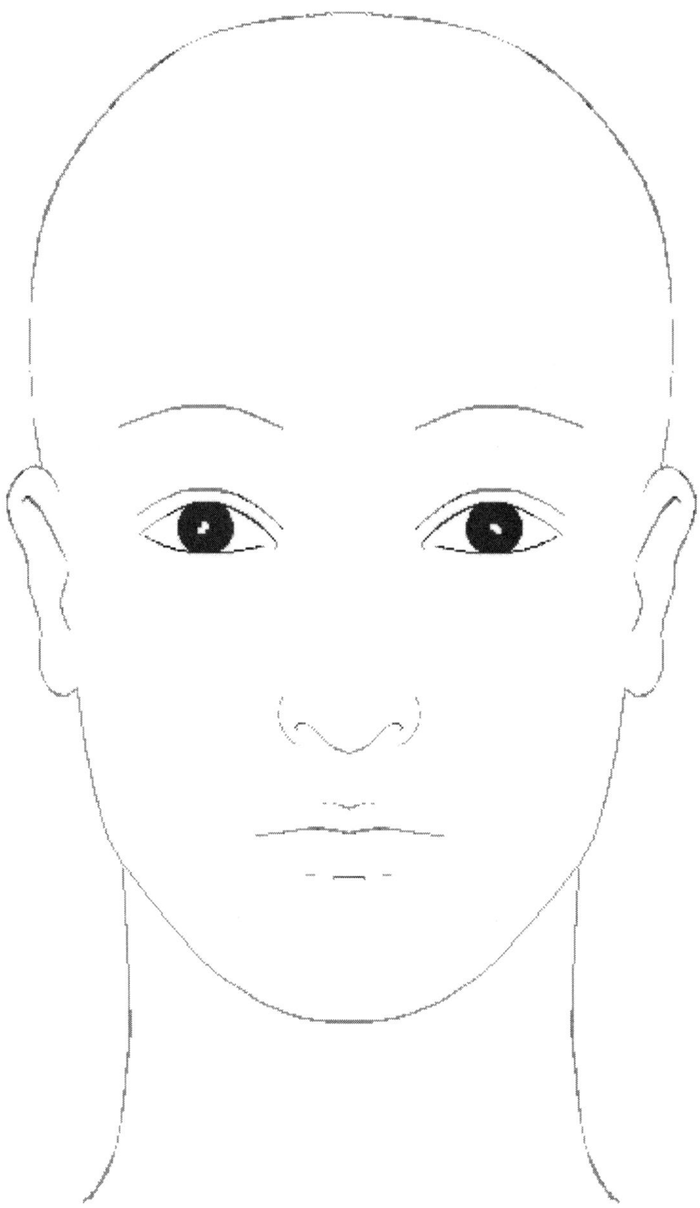

3. 분장 디자인 - 측면 Makeup Design - Side View

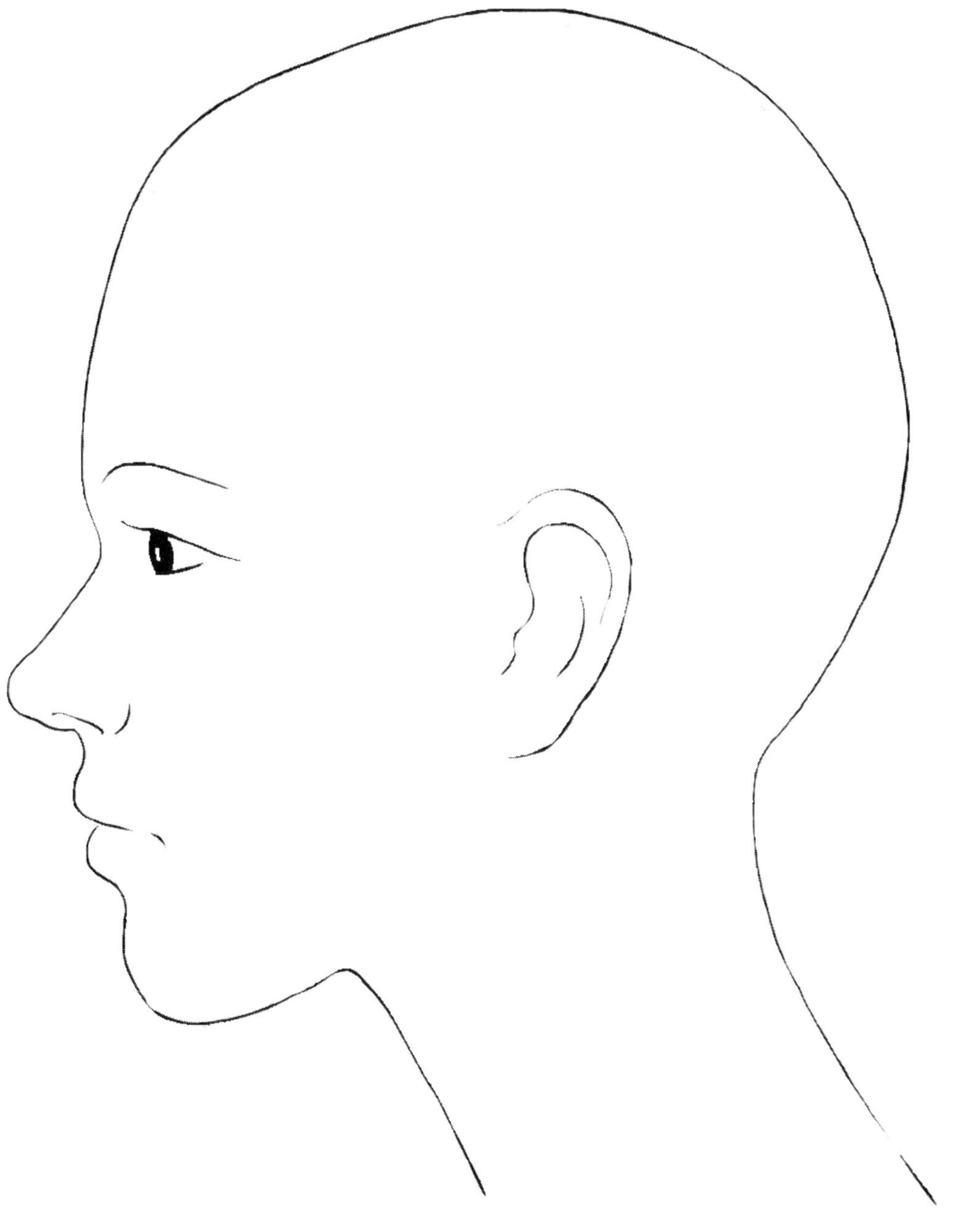

4. 스트레이트 분장 완성 - 정면 Straight Makeup Completed - Front View

스트레이트 분장 완성 정면 이미지 삽입
Insert Straight Makeup Completed Front View Image.

4. 스트레이트 분장 완성 - 측면 Straight Makeup Completed - Side View

스트레이트 분장 완성 측면 이미지 삽입
Insert Straight Makeup Completed Side View Image.

5. 분장 계획 시행 Implementation of Makeup Plan

분장 계획 시행

Work Name 작품명		Actor 배우	

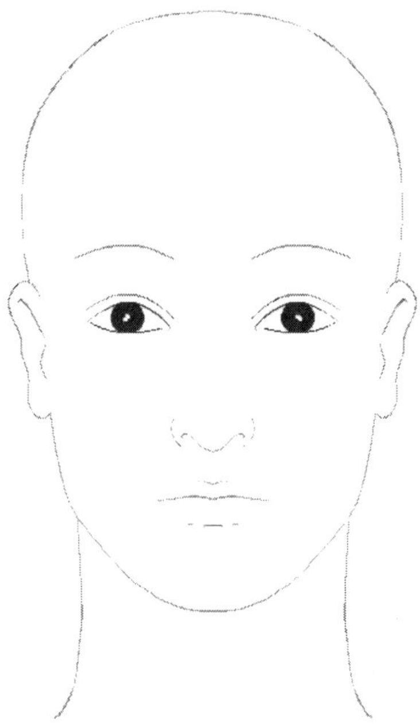

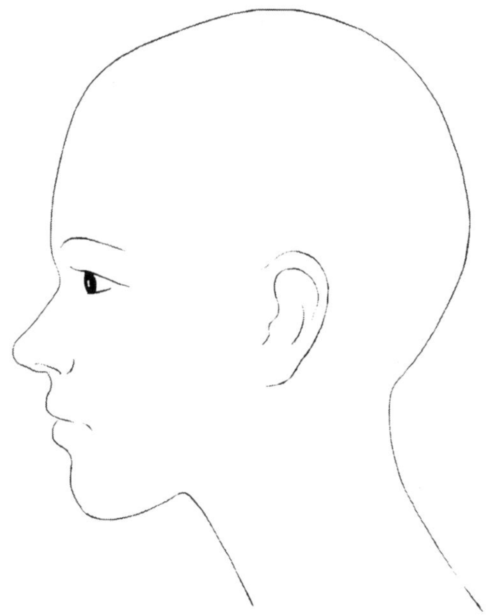

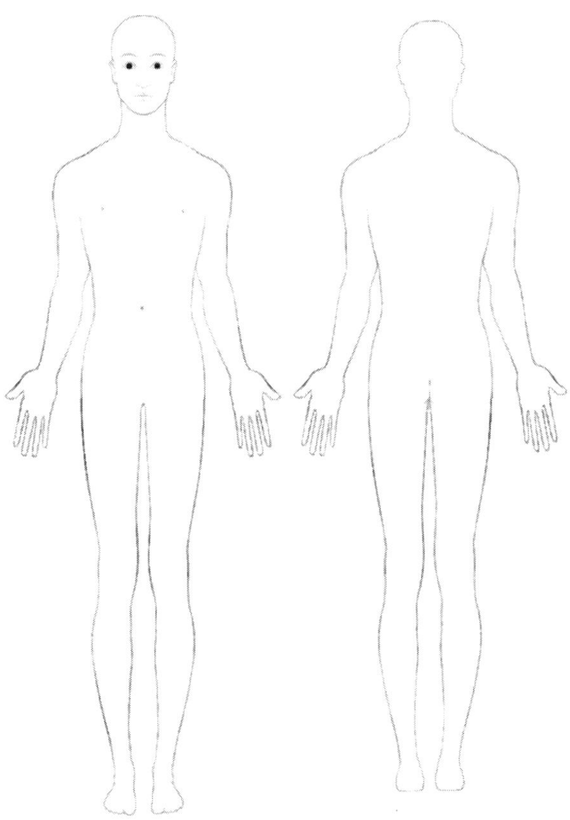

이마 Forehead	아이라인 Eye liners	눈 Eye
코 Nose	뺨 Cheek	입술 Mouth
베이스 Base	섀도우 Shadow	파우더 Powder
헤어 Hair	메모 Note	

6. 등장인물 분석 Character Analysis

등장인물 분석	
작품명 Work Name	장르 Genre
배역명 Character Name	역할 Part
나이/성별 Age/Sex	성격 Personality
건강 Health	환경 Environment
유전학 Genetics	시대/국적 Times/Nationality
특징 Characteristic	

7. 분장의도 및 성격표현 Expression of Makeup Design and Character Portrayal

분장의도 및 성격표현		
	설명 Explain	만족도 Satisfaction
분장 의도 Makeup Purpose		
성격 표현 Character Expression		

Lesson 4

노인 분장
Aging Makeup

Lesson 4. 노인 분장 Aging Makeup

1. 연령별 노화의 특징 Characteristics of Aging by Age

연 령 Age		내 용 Content
10대	피부의 특징 Characteristics of the Skin	
	피부의 변화 Changes in the Skin	
20대	피부의 특징 Characteristics of the Skin	
	피부의 변화 Changes in the Skin	
30대	피부의 특징 Characteristics of the Skin	
	피부의 변화 Changes in the Skin	
40대	피부의 특징 Characteristics of the Skin	
	피부의 변화 Changes in the Skin	
50대	피부의 특징 Characteristics of the Skin	
	피부의 변화 Changes in the Skin	
60대	피부의 특징 Characteristics of the Skin	
	피부의 변화 Changes in the Skin	
70대	피부의 특징 Characteristics of the Skin	
	피부의 변화 Changes in the Skin	
80대 이후	피부의 특징 Characteristics of the Skin	
	피부의 변화 Changes in the Skin	

2. 자료조사 Research

10대 얼굴 이미지 삽입
Insert an Image of a Face in Their 10s

20대 얼굴 이미지 삽입
Insert an Image of a Face in Their 20s

30대 얼굴 이미지 삽입
Insert an Image of a Face in Their 30s

40대 얼굴 이미지 삽입
Insert an Image of a Face in Their 40s

50대 얼굴 이미지 삽입
Insert an Image of a Face in Their 50s

60대 얼굴 이미지 삽입
Insert an Image of a Face in Their 60s

70대 얼굴 이미지 삽입
Insert an Image of a Face in Their 70s

80대 얼굴 이미지 삽입
Insert an Image of a Face in Their 80s

3. 분장 디자인 - 정면 Makeup Design - Front View

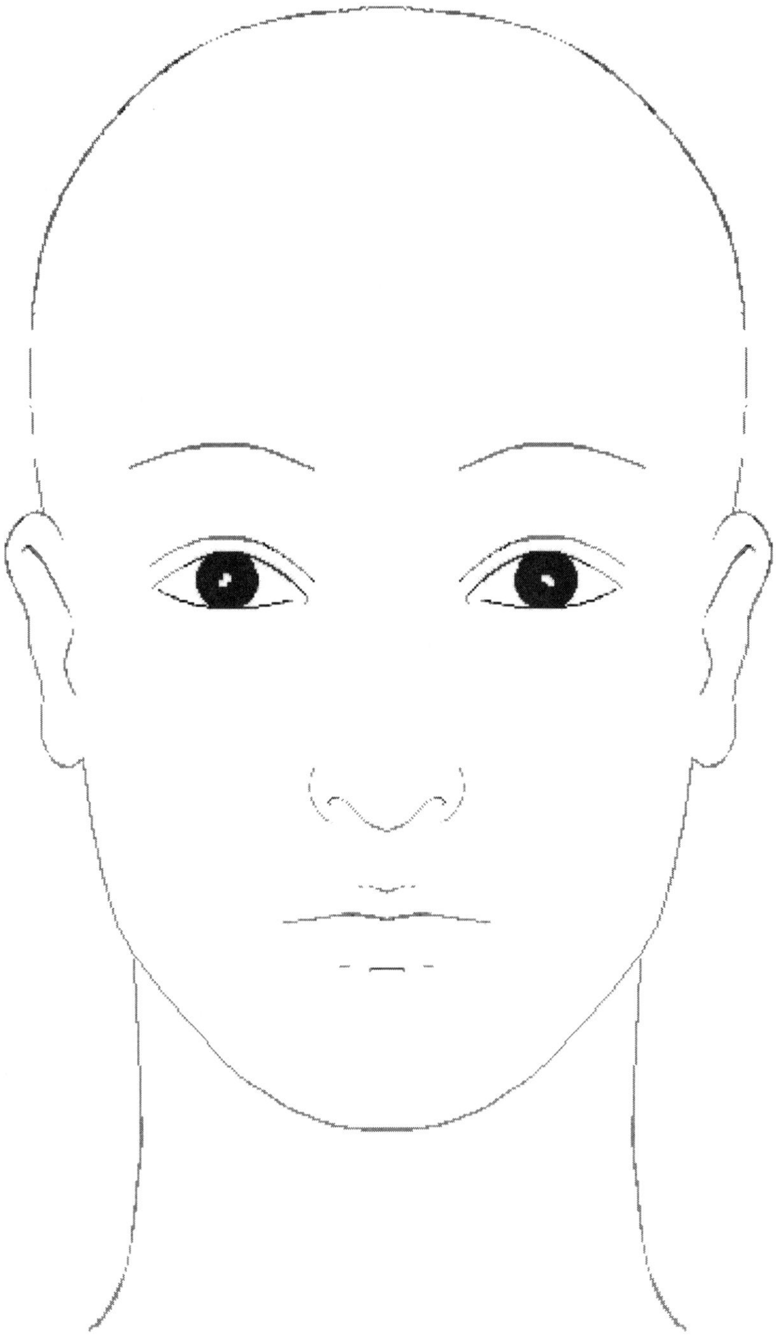

3. 분장 디자인 - 측면 Makeup Design - Side View

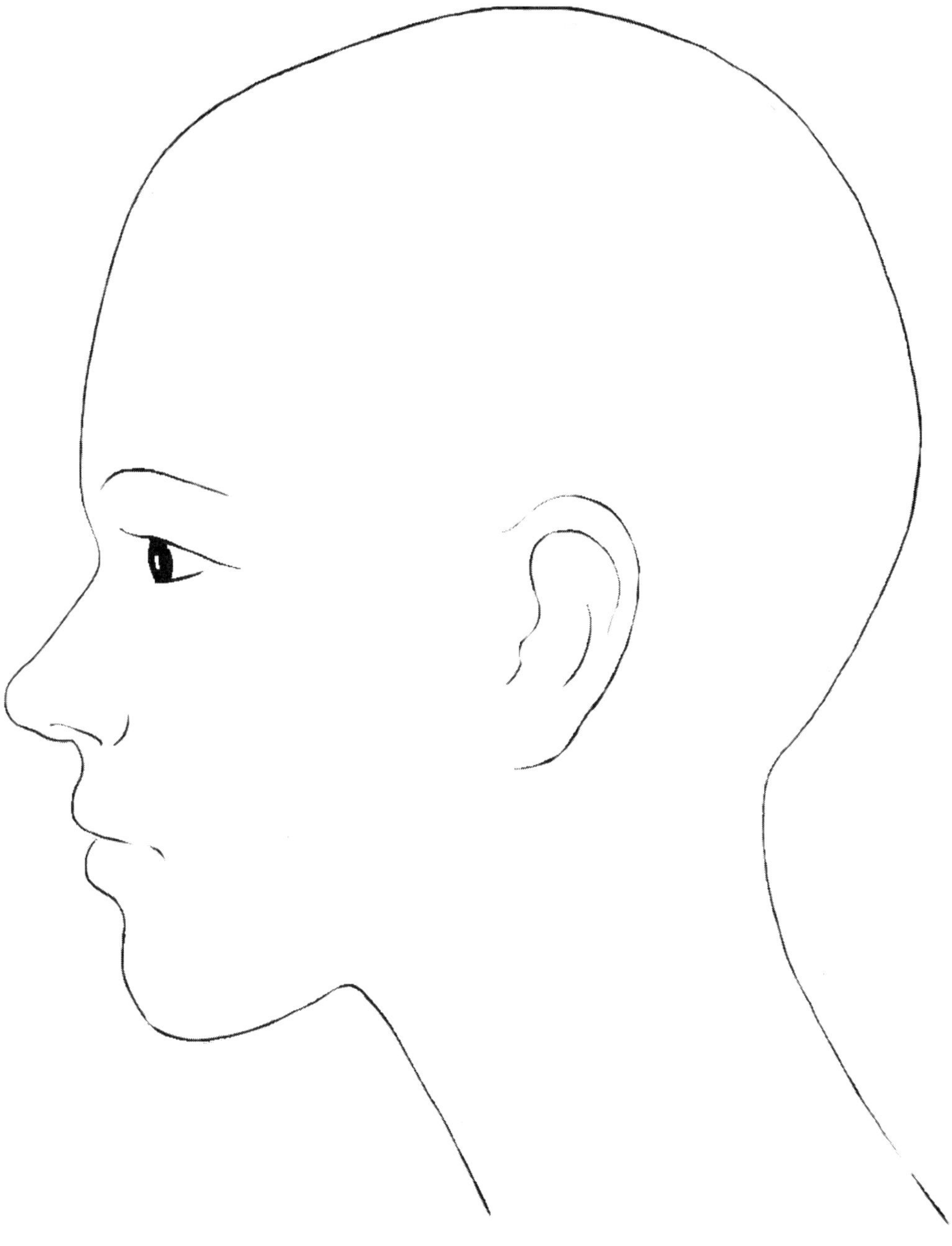

4. 노인 분장 완성 - 정면 Elderly Makeup Completed - Front View

노인 분장 완성 정면 이미지 삽입
Insert a Completed Front View Image of Elderly Makeup.

4. 노인 분장 완성 - 측면 Elderly Makeup Completed - Side View

노인 분장 완성 측면 이미지 삽입
Insert a Completed Side View Image of Elderly Makeup.

5. 분장 계획 시행 Implementation of Makeup Plan

분장 계획 시행

Work Name 작품명		Actor 배우	

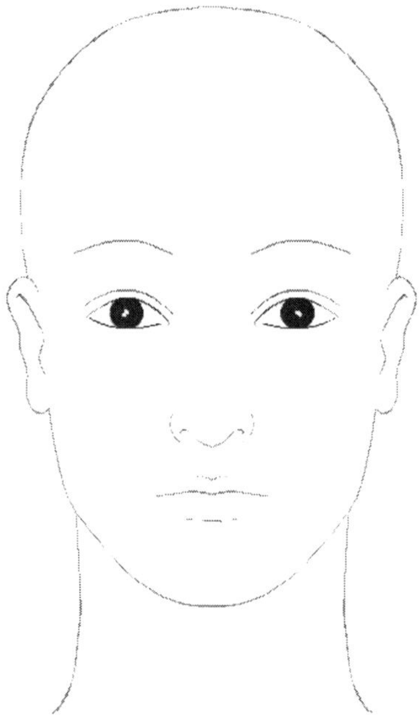

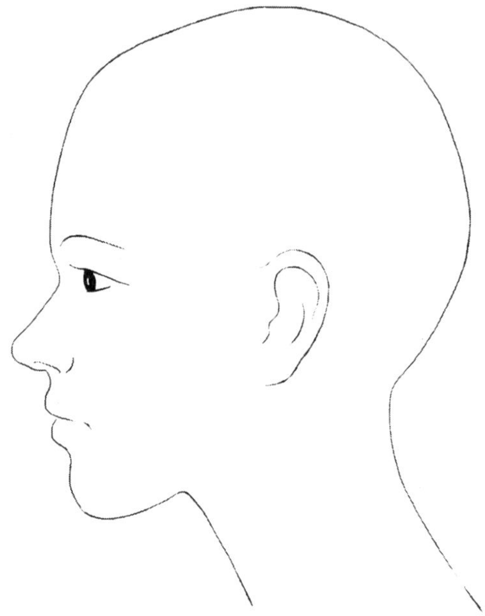

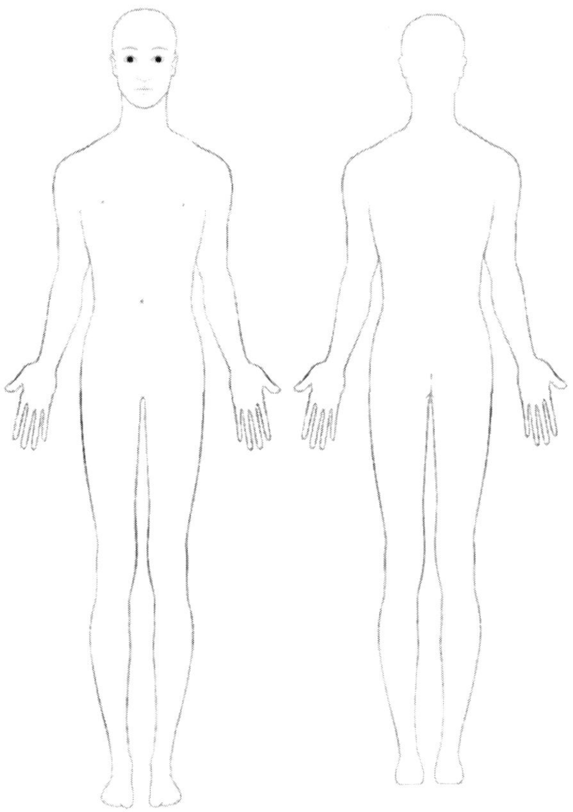

이마 Forehead	아이라인 Eye liners	눈 Eye
코 Nose	뺨 Cheek	입술 Mouth
베이스 Base	샤도우 Shadow	파우더 Powder
헤어 Hair	메모 Note	

6. 등장인물 분석 Character Analysis

등장인물 분석	
작품명 Work Name	장르 Genre
배역명 Character Name	역할 Part
나이/성별 Age/Sex	성격 Personality
건강 Health	환경 Environment
유전학 Genetics	시대/국적 Times/Nationality
특징 Characteristic	

7. 분장의도 및 성격표현 Expression of Makeup Design and Character Portrayal

분장의도 및 성격표현		
	설명 Explain	만족도 Satisfaction
분장 의도 Makeup Purpose		
성격 표현 Character Expression		

Lesson 5

상처 분장1
Wound Makeup1

Lesson 5. 상처 분장1 Wound Makeup1

1. 인조 피 제작 Fake Blood Production

□ 준비물 Materials

-
-
-
-
-

준비물 이미지 삽입
Insert Materials Image.

□ 인조 피의 종류 Types of Fake Blood

-
-
-
-
-

2. 타박상 분장 Bruise Makeup

1) 타박상이란? What is a Bruise?

2) 타박상 분장 재료 Bruise Makeup Materials

타박상 분장 재료 이미지 삽입
Insert Bruise Makeup Materials Image.

3) 타박상 자료조사 Bruise Research

타박상 자료수집 이미지 삽입
Insert Bruise Data Research Image.

4) 타박상 분장 디자인 Bruise Makeup Design

5) 타박상 분장 완성 Bruise Makeup Completed

■ 타박상의 원인 Causes of Bruises :

타박상 분장 완성 이미지 삽입 Insert Bruise Makeup Completed Image.

3. 찰과상 분장 Abrasion Makeup

1) 찰과상이란? What is an Abrasion?

2) 찰과상 분장 재료 Abrasion Makeup Materials

찰과상 분장 재료 이미지 삽입
Insert Abrasion Makeup Materials Image.

3) 찰과상 자료조사 Abrasion Research

찰과상 자료수집 이미지 삽입

Insert Abrasion Data Research Image.

4) 찰과상 분장 디자인 Abrasion Makeup Design

5) 찰과상 분장 완성 Abrasion Makeup Completed

■ 찰과상의 원인 Causes of Abrasions :

찰과상 분장 완성 이미지 삽입
Insert Abrasion Makeup Completed Image.

4. 열상 분장 Laceration Makeup

1) 열상이란? What is an Laceration?

2) 열상 분장 재료 Laceration Makeup Materials

열상 분장 재료 이미지 삽입
Insert Laceration Makeup Materials Image.

3) 열상 자료조사 Laceration Research

열상 자료수집 이미지 삽입
Insert Laceration Data Collection Image.

4) 열상 분장 디자인 Laceration Makeup Design

5) 열상 분장 완성 Laceration Makeup Completed

■ 열상의 원인 Causes of Lacerations :

타박상 분장 완성 이미지 삽입
Insert Laceration Makeup Completed Image.

5. 분장 디자인 적용부위 Makeup Design Application Area

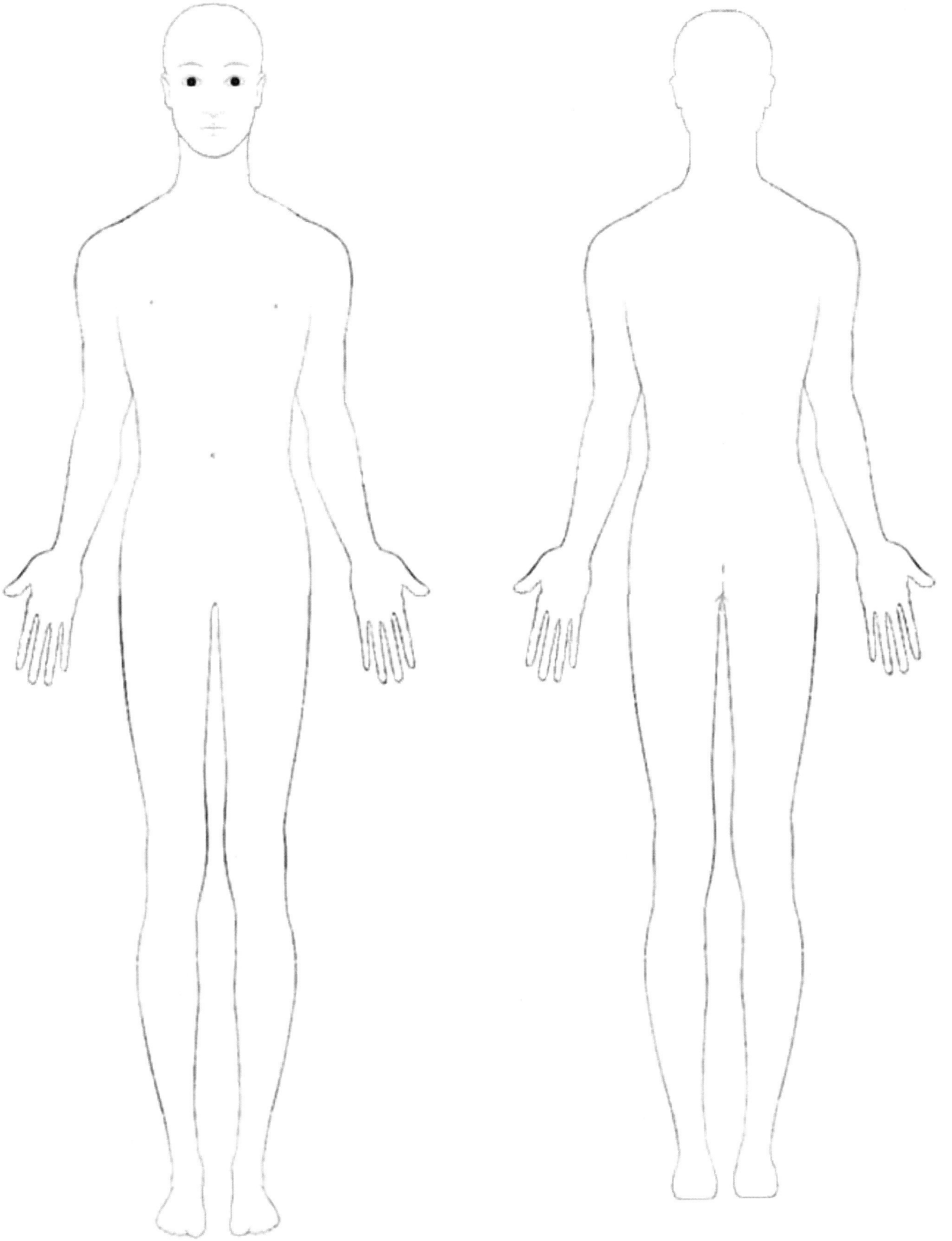

6. 분장 계획 시행 Implementation of Makeup Plan

분장 계획 시행			
Work Name 작품명		Actor 배우	

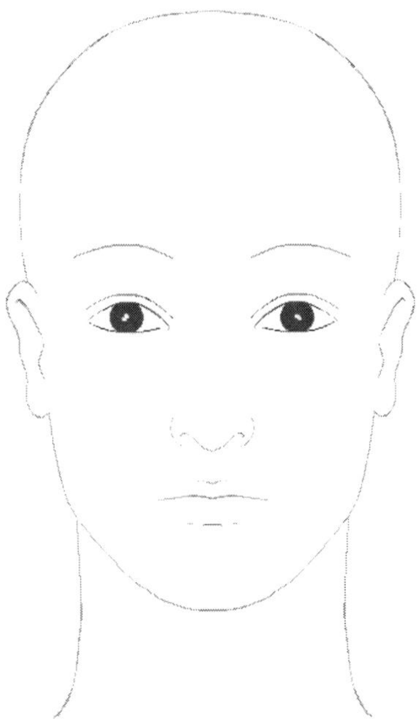

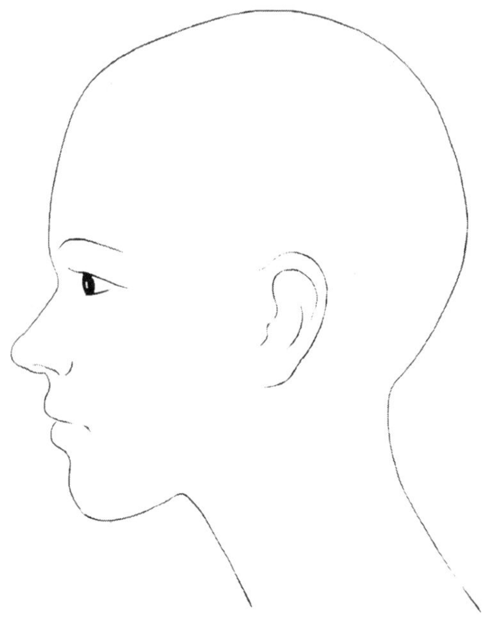

이마 Forehead	눈썹 Eyebrows	눈 Eye	코 Nose	뺨 Cheek
입 Mouth	화운데이션 Foundation	아이라인 Eye liners	가장 밝은 부분 Highlight	명암 Shading
파우더 Powder	목, 어깨 Neck, Shoulder	팔 Arm	복부. 배 Abdomen	등 Back
다리 Leg	손, 손가락 Hand, Finger	발, 발가락 Foot, Toe	헤어 Hair	그 외 메모사항 Other Notes

7. 등장인물 분석 Character Analysis

등장인물 분석	
작품명 Work Name	장르 Genre
배역명 Character Name	역할 Part
나이/성별 Age/Sex	성격 Personality
건강 Health	환경 Environment
유전학 Genetics	시대/국적 Times/Nationality
특징 Characteristic	

8. 분장의도 및 성격표현 Expression of Makeup Design and Character Portrayal

분장의도 및 성격표현		
	설명 Explain	만족도 Satisfaction
분장 의도 Makeup Purpose		
성격 표현 Character Expression		

Lesson 6

상처 분장2
Wound Makeup2

Lesson 6. 상처 분장2 Wound Makeup2

1. 봉합 상처 분장 Sutured Wound Makeup

1) 봉합 상처란? What is a Sutured Wound?

2) 봉합 상처 분장 재료 Sutured Wound Makeup Materials

봉합상 분장 재료 이미지 삽입
Insert Sutured Wound Makeup Materials Image.

3) 봉합 상처 자료조사 Sutured Wound Research

봉합상처 자료수집 이미지 삽입
Insert Sutured Wound Data Collection Image.

4) 봉합 상처 분장 디자인 Sutured Wound Makeup Design

5) 봉합 상처 분장 완성 Sutured Wound Makeup Completed

■ 봉합 상처의 원인 Causes of Sutured Wounds :

봉합상처 분장 완성 이미지 삽입
Insert Sutured Wound Makeup Completed Image.

2. 창상 분장 Wound Makeup

1) 창상이란? What is a Wound?

2) 창상 분장 재료 Wound Makeup Materials

창상 분장 재료 이미지 삽입
Insert Wound Makeup Materials Image.

3) 창상 자료조사 Wound Research

창상 자료수집 이미지 삽입

Insert Wound Data Collection Image.

4) 창상 분장 디자인 Wound Makeup Design

5) 창상 분장 완성 Wound Makeup Completed

■ 창상의 원인 Causes of Wounds :

창상 분장 완성 이미지 삽입
Insert Wound Makeup Completed Image.

3. 총상 분장 Gunshot Wound Makeup

1) 총상이란? What is a Gunshot Wound?

2) 총상 분장 재료 Gunshot Makeup Materials

총상 분장 재료 이미지 삽입
Insert Gunshot Wound Makeup Materials Image.

3) 총상 자료조사 Gunshot Wound Research

총상 자료수집 이미지 삽입
Insert Gunshot Wound Data Collection Image.

4) 총상 분장 디자인 Gunshot Wound Makeup Design

5) 총상 분장 완성 Gunshot Wound Makeup Completed

■ 총상의 원인 Causes of Gunshot Wounds :

총상 분장 완성 이미지 삽입
Insert Gunshot Wound Makeup Completed Image.

4. 자상 분장 Stab Wound Makeup

1) 자상이란? What is a Stab Wound?

2) 자상 분장 재료 Stab Makeup Materials

자상 분장 재료 이미지 삽입
Insert Stab Wound Makeup Materials Image.

3) 자상 자료조사 Stab Wound Research

자상 자료수집 이미지 삽입
Insert Stab Wound Data Collection Image.

4) 자상 분장 디자인 Stab Wound Makeup Design

5) 자상 분장 완성 Stab Wound Makeup Completed

■ 자상의 원인 Causes of Stab Wounds :

자상 분장 완성 이미지 삽입
Insert Stab Wound Makeup Completed Image.

5. 재해 분장(파편이 박힌 상처) Disaster Makeup(Wound with Embedded Debris)

1) 재해란? What is a Disaster?

- 자연재해 Man-Made Disaster

- 인재 Man-Made Disaster

2) 재해 분장 재료 Disaster Makeup Materials

자상 분장 재료 이미지 삽입
Attech Disaster Makeup Materials Image.

3) 재해 자료조사 Disaster Research

재해 자료수집 이미지 삽입
Insert Disaster Data Collection Image.

4) 재해 분장 디자인 Disaster Makeup Design

5) 재해 분장 완성 Disaster Makeup Completed

■ 재해의 원인 Cause of the Disaster :

재해 분장 완성 이미지 삽입
Insert Disaster Makeup Completed Image.

6. 분장 디자인 적용부위 Makeup Design Application Area

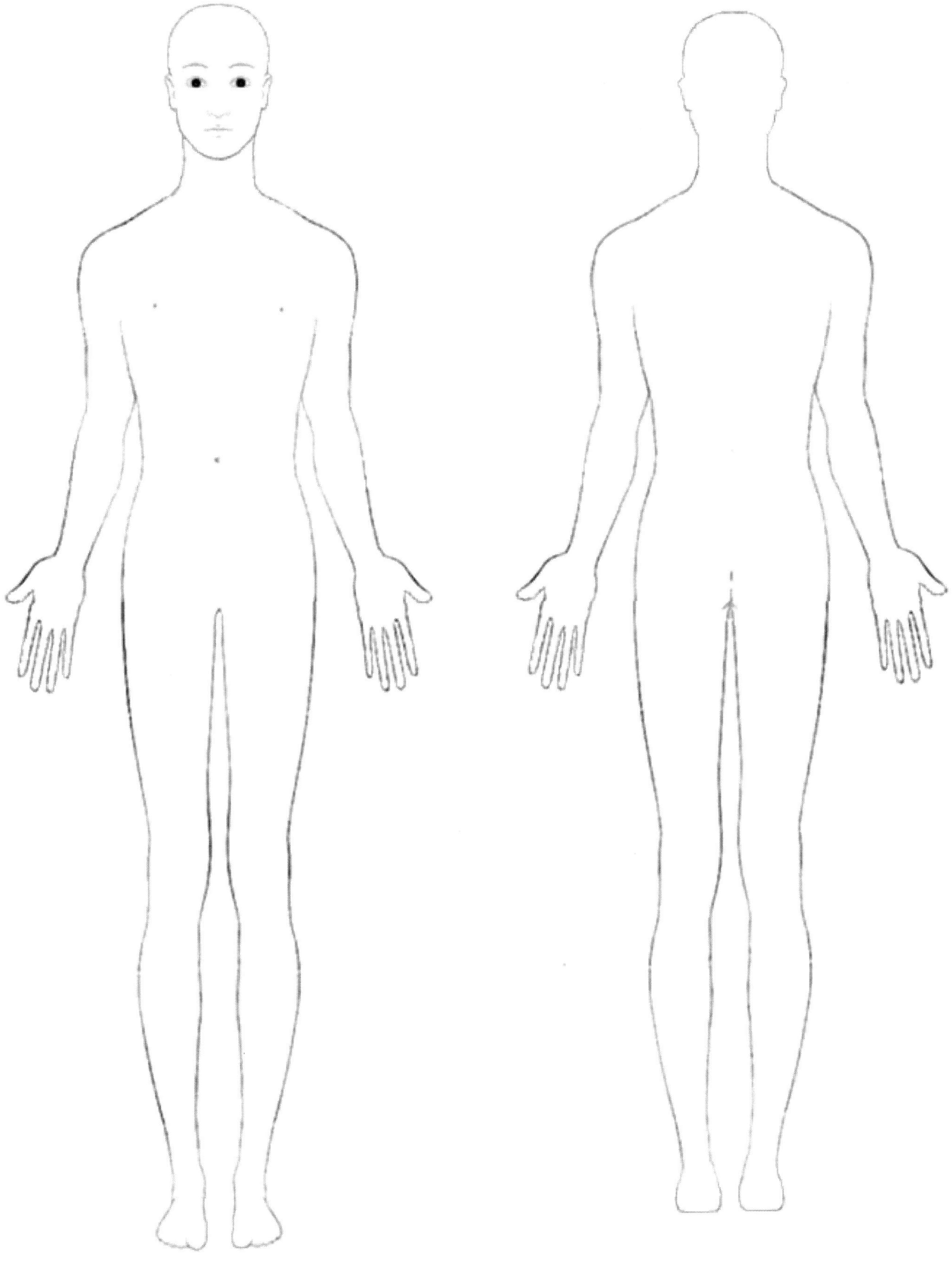

7. 분장 계획 시행 Implementation of Makeup Plan

분장 계획 시행			
Work Name 작품명		Actor 배우	

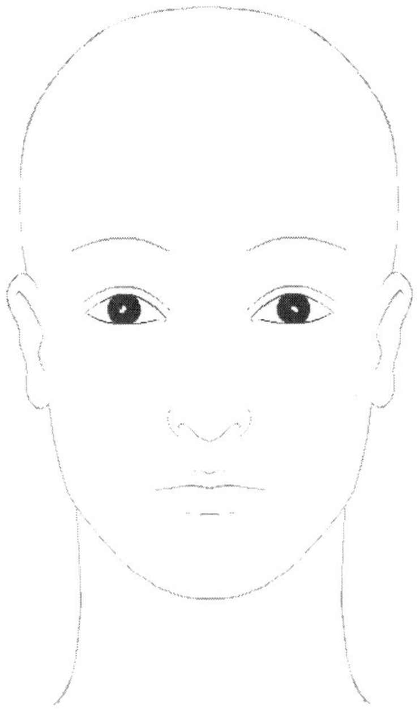

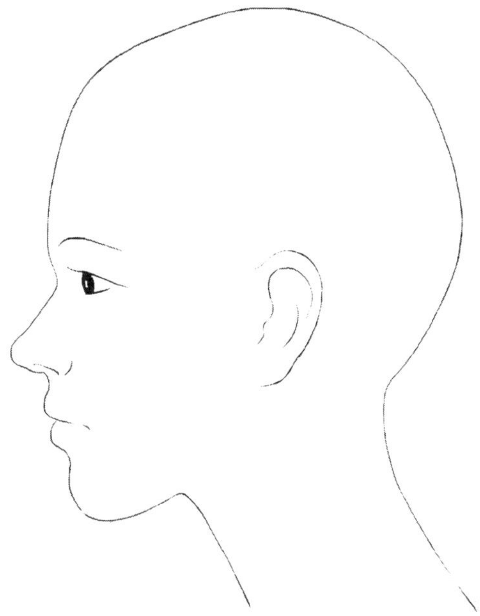

이마 Forehead	눈썹 Eyebrows	눈 Eye	코 Nose	뺨 Cheek
입 Mouth	화운데이션 Foundation	아이라인 Eye liners	가장 밝은 부분 Highlight	명암 Shading
파우더 Powder	목, 어깨 Neck, Shoulder	팔 Arm	복부. 배 Abdomen	등 Back
다리 Leg	손, 손가락 Hand, Finger	발, 발가락 Foot, Toe	헤어 Hair	그 외 메모사항 Other Notes

8. 등장인물 분석 Character Analysis

등장인물 분석	
작품명 Work Name	장르 Genre
배역명 Character Name	역할 Part
나이/성별 Age/Sex	성격 Personality
건강 Health	환경 Environment
유전학 Genetics	시대/국적 Times/Nationality
특징 Characteristic	

9. 분장의도 및 성격표현 Expression of Makeup Design and Character Portrayal

분장의도 및 성격표현		
	설명 Explain	만족도 Satisfaction
분장 의도 Makeup Purpose		
성격 표현 Character Expression		

Lesson 7

대머리 분장
Bald Makeup

Lesson 7. 대머리 분장 Bald Makeup

1. 대머리 분장

1) 대머리 Bald

2) 대머리 자료조사 Bald Research

대머리 분장 자료수집 이미지 삽입
Bald Data Collection Image Included.

2. 볼드캡 제작 Bald Cap Production

- 준비물 Materials

볼드캡 제작 과정
이미지 삽입1
Insert Bald Cap Production
Process Image1.

⇨

볼드캡 제작 과정
이미지 삽입2
Insert Bald Cap Production
Process Image2.

⇨

볼드캡 제작 과정
이미지 삽입3
Insert Bald Cap Production
Process Image3.

3. 볼드캡 씌우기 및 제거 Bald Cap Application and Removal

1) 볼드캡 씌우기 과정 Bald Cap Application Process

볼드캡 씌우기 과정 이미지 삽입1 Insert Bald Cap Application Process Image1.	⇨	볼드캡 씌우기 과정 이미지 삽입2 Insert Bald Cap Application Process Image2.	⇨	볼드캡 씌우기 과정 이미지 삽입3 Insert Bald Cap Application Process Image3.

2) 볼드캡 제거 Bald Cap Removal

볼드캡 제거 과정 이미지 삽입1 Insert Bald Cap Removal Process Image1.	⇨	볼드캡 제거 과정 이미지 삽입2 Insert Bald Cap Removal Process Image2.	⇨	볼드캡 제거 과정 이미지 삽입3 Insert Bald Cap Removal Process Image3.

4. 분장 디자인 - 정면 Makeup Design - Front View

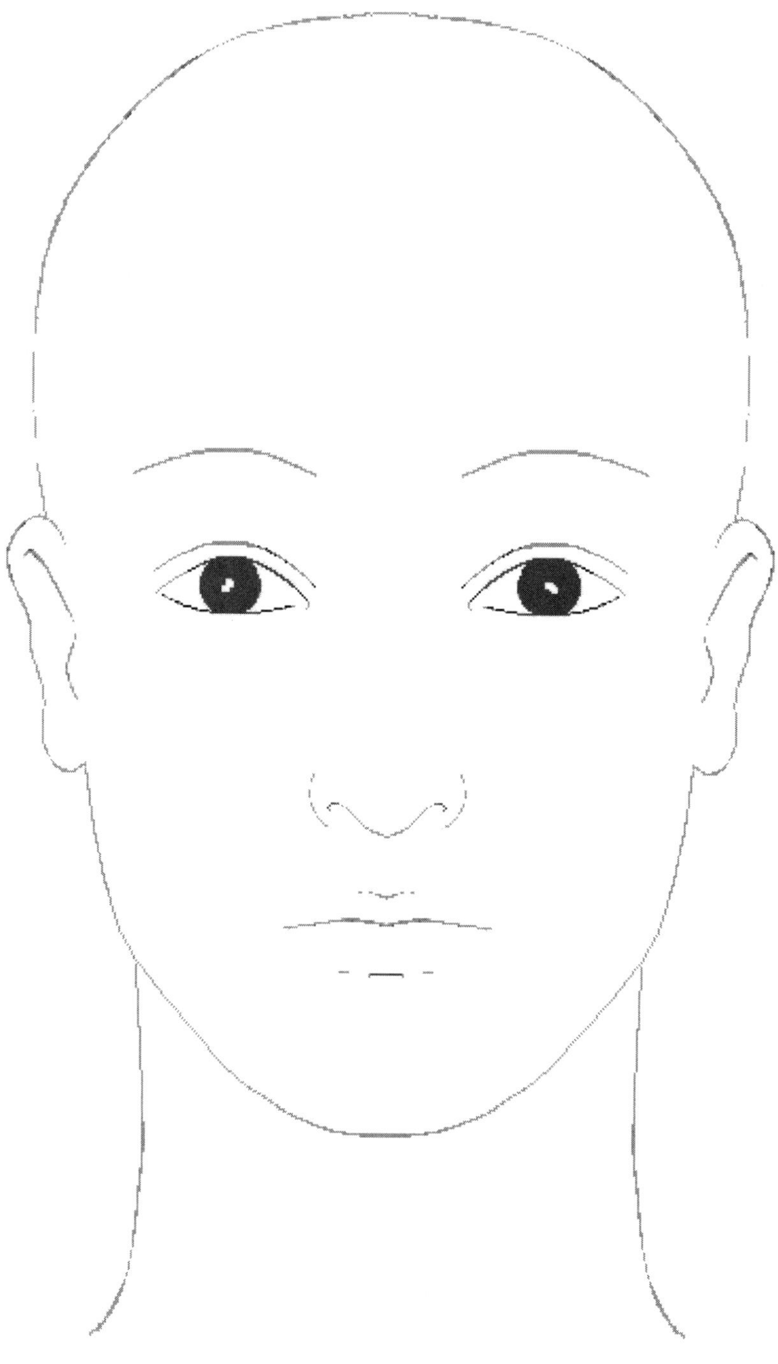

4. 분장 디자인 - 측면 Makeup Design - Side View

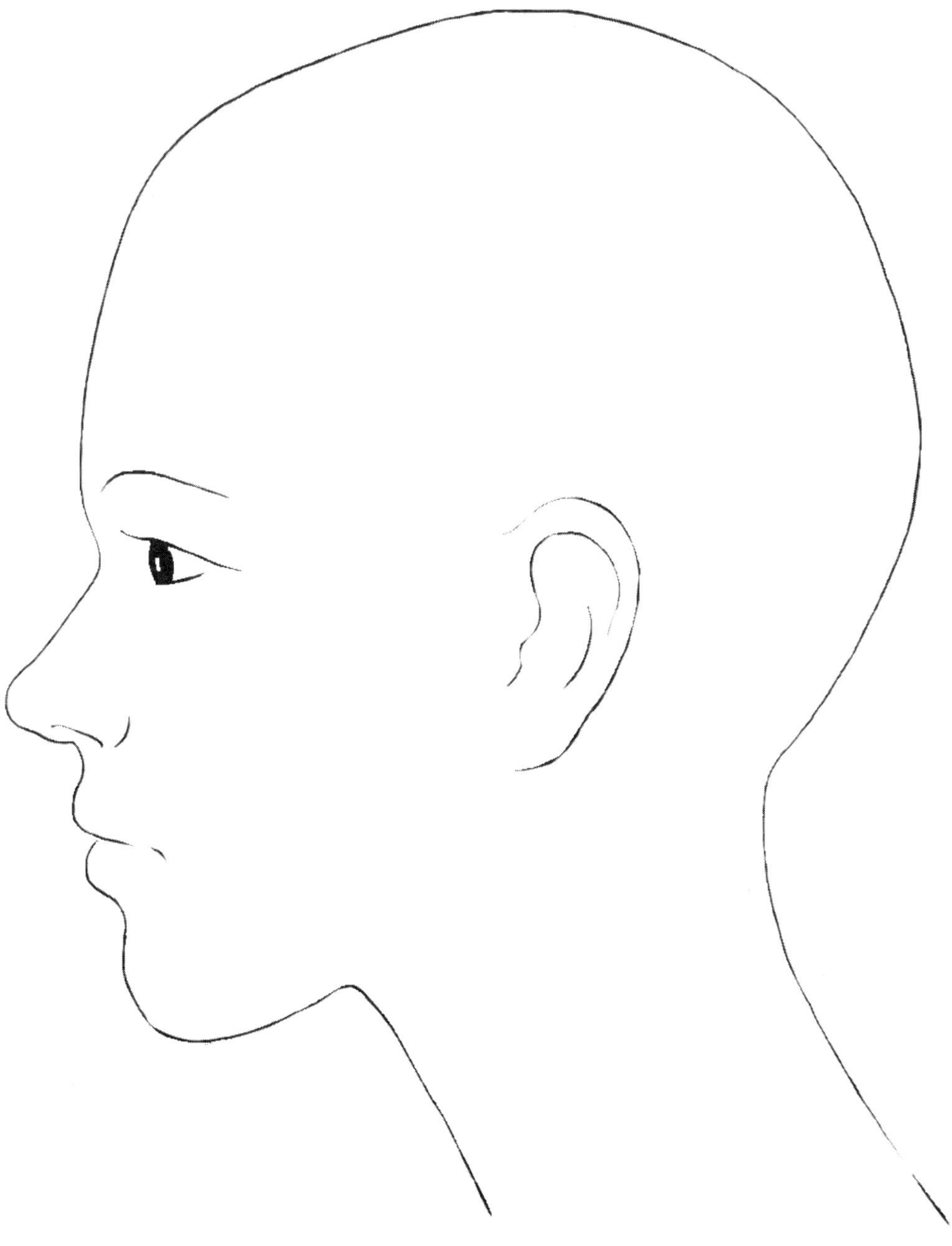

5. 대머리 분장 완성 - 정면 Bald Makeup - Front View

완성 분장 정면 이미지 삽입

Insert Completed Makeup Front View Image.

5. 대머리 분장 완성 - 측면, 뒷면 Bald Makeup - Side View, Back View

완성 분장 좌측면 이미지 삽입 Insert Completed Makeup left Side View Image.	완성 분장 우측면 이미지 삽입 Insert Completed Makeup Right Side View Image.
대머리 분장 뒷면 이미지 삽입 Insert Images of Bald Makeup From the Back.	

6. 분장 계획 시행 Implementation of Makeup Plan

분장 계획 시행			
Work Name 작품명		Actor 배우	

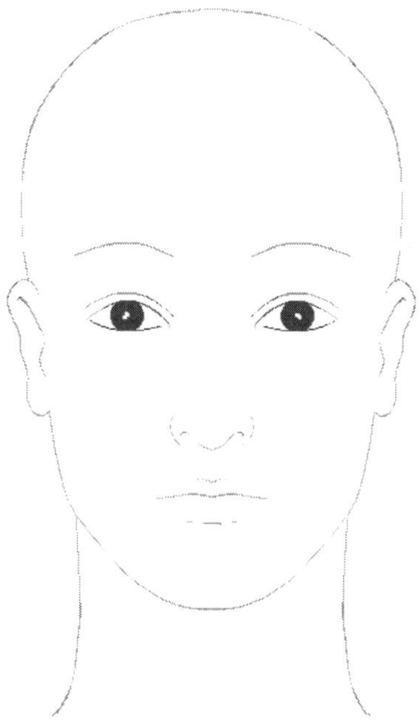

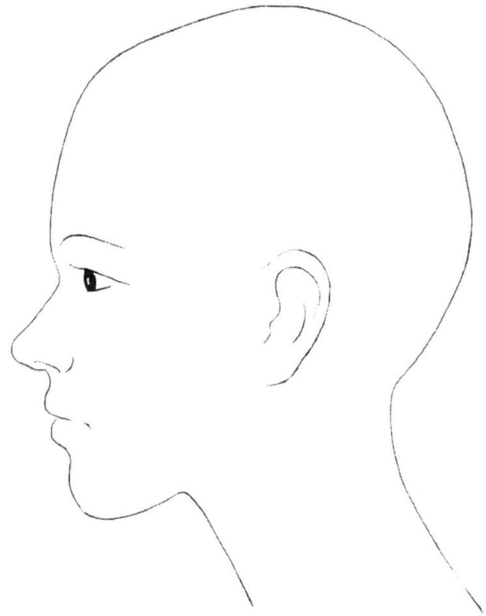

이마 Forehead	눈썹 Eyebrows	아이라인 Eyeline
코 Nose	뺨 Cheek	입 Mouth
헤어라인 측면 Side Hairline	뒷목 Back of the Neck	화운데이션 Foundation
파우더 Powder	메모 Notes	

7. 등장인물 분석 Character Analysis

등장인물 분석	
작품명 Work Name	장르 Genre
배역명 Character Name	역할 Part
나이/성별 Age/Sex	성격 Personality
건강 Health	환경 Environment
유전학 Genetics	시대/국적 Times/Nationality
특징 Characteristic	

8. 분장의도 및 성격표현 Expression of Makeup Design and Character Portrayal

분장의도 및 성격표현		
	설명 Explain	만족도 Satisfaction
분장 의도 Makeup Purpose		
성격 표현 Character Expression		

Lesson 8

수염 분장
Beard Makeup

Lesson 8. 수염 분장 Beard Makeup

1. 수염 분장 Beard Makeup

1) 콧수염 Mustache

2) 턱수염 Goatee

3) 구레나룻 수염 SideBurns

4) 수염 분장 재료 Beard Makeup Materials

수염 분장 재료 이미지 삽입
Insert Beard Makeup Materials Image.

2. 자료조사 Research

수염 분장 자료 이미지 삽입

Insert beard Makeup reference Image.

3. 분장 디자인 - 정면 Makeup Design - Front View

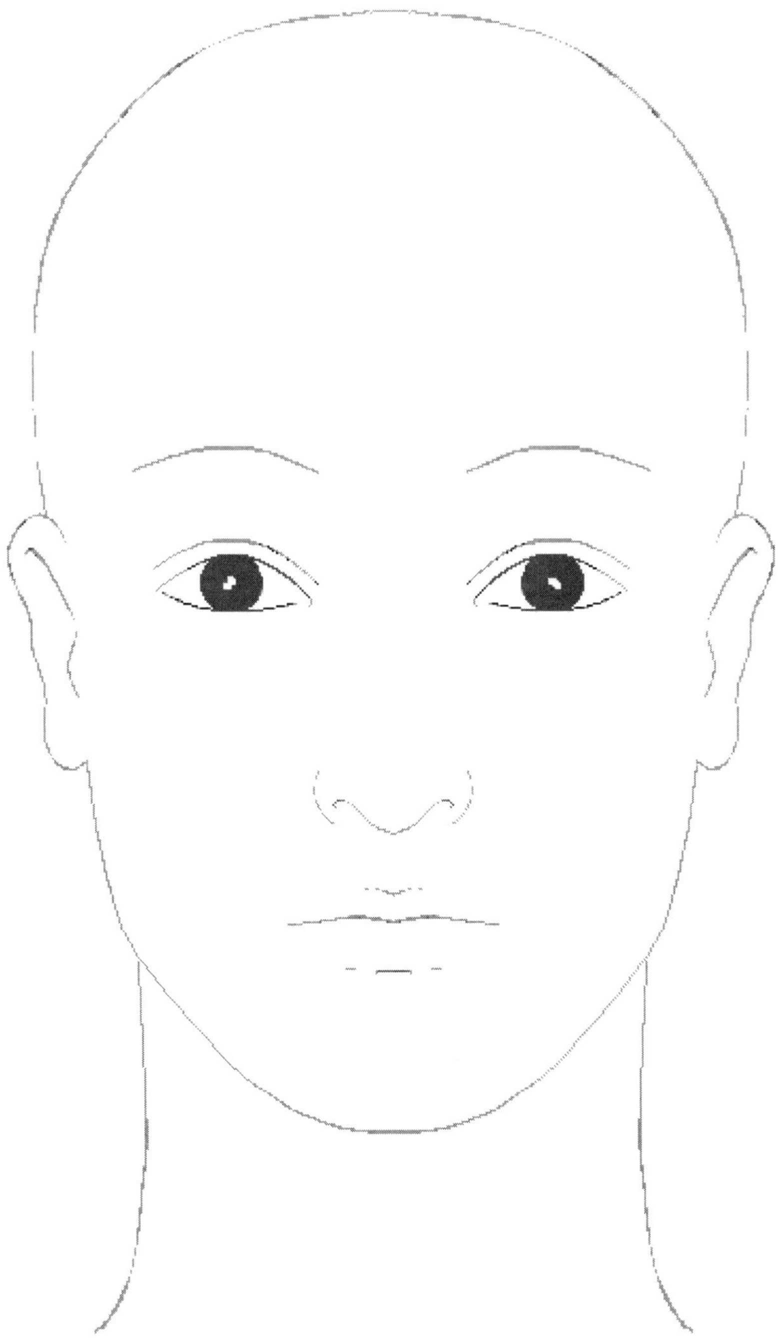

3. 분장 디자인 - 측면 Makeup Design - Side View

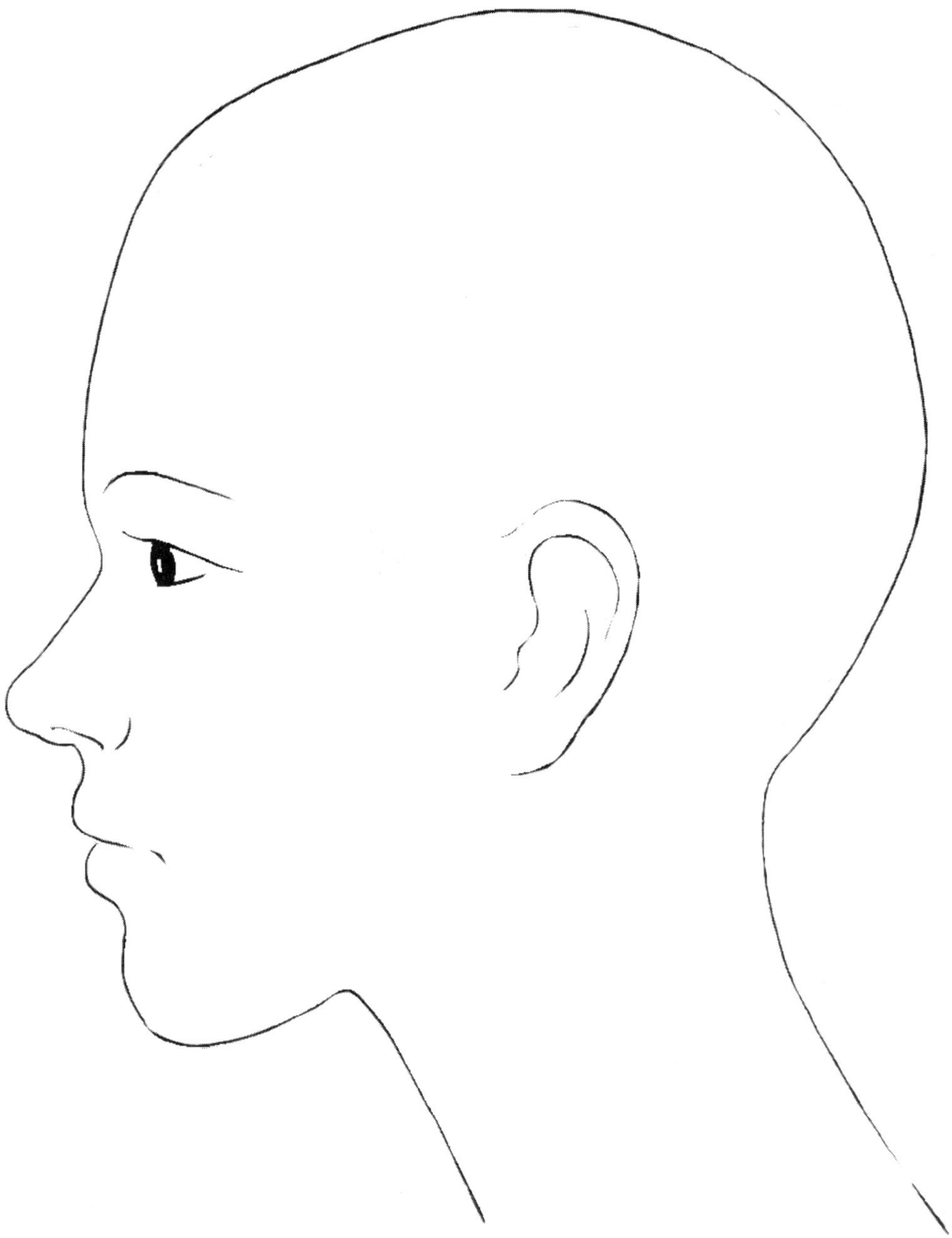

4. 수염 분장 완성 - 정면 Beard Makeup Completed - Front View

분장 완성 정면 이미지 삽입
Insert Completed Makeup Front View Image.

4. 수염 분장 완성 - 측면 Beard Makeup Completed - Side View

분장 완성 측면 이미지 삽입
Insert Completed Makeup Side View Image.

5. 분장 계획 시행 Implementation of Makeup Plan

분장 계획 시행

Work Name 작품명		Actor 배우	

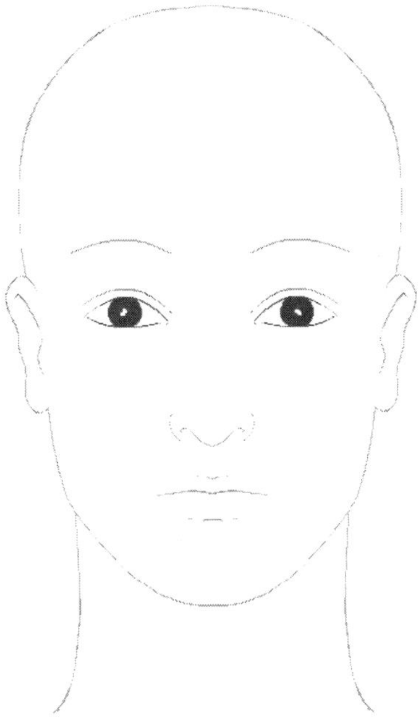

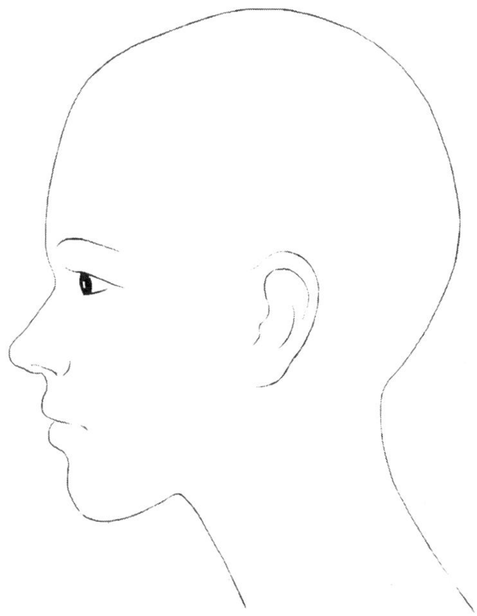

이마 Forehead	눈썹 Eyebrows	눈 Eye	코 Nose	뺨 Cheek
입 Mouth	화운데이션 Foundation	아이라인 Eye liners	가장 밝은 부분 Highlight	명암 Shading
파우더 Powder	목, 어깨 Neck, Shoulder	귀밑 Below the Ear	코밑 Below the Nose	입밑 Below the Mouth
헤어 Hair	그 외 메모사항 Other Notes			

6. 등장인물 분석 Character Analysis

등장인물 분석	
작품명 Work Name	장르 Genre
배역명 Character Name	역할 Part
나이/성별 Age/Sex	성격 Personality
건강 Health	환경 Environment
유전학 Genetics	시대/국적 Times/Nationality
특징 Characteristic	

7. 분장의도 및 성격표현 Expression of Makeup Design and Character Portrayal

분장의도 및 성격표현		
	설명 Explain	만족도 Satisfaction
분장 의도 Makeup Purpose		
성격 표현 Character Expression		

Lesson 9

동상 분장
Frostbite Makeup

Lesson 9. 동상 분장 Frostbite Makeup

1. 자료조사 Research

• **동상 분장 재료 Frostbite Makeup Materials**

동상 분장 재료 이미지 삽입
Insert Frostbite Makeup Materials Image.

2. 동상 Frostbite

- 동상이란 신체의 피부가 극심한 추위에 노출되어 피부 조직이 상해를 입는 것이다.
- 동상은 대개 심장으로부터 멀리 떨어진 부위, 특히 바깥에 많이 노출되는 손, 발 등의 부위에서 잘 일어난다.

1) 1도 동상 First-Degree Frostbite

- **특징**
- **증상**
- **치료**

동상 관련 이미지 삽입
Insert Images Related to Frostbite.

2) 2도 동상 Second-Degree Frostbite

- **특징**
- **증상**
- **치료**

동상 관련 이미지 삽입
Insert Images Related to Frostbite.

3) 3도 동상 Third-Degree Frostbite

- **특징**
- **증상**
- **치료**

동상 관련 이미지 삽입
Insert Images Related to Frostbite.

4) 4도 동상 Fourth-Degree Frostbite

- **특징**
- **증상**
- **치료**

동상 관련 이미지 삽입
Insert Images Related to Frostbite.

3. 분장 디자인 부위 Makeup Design Area

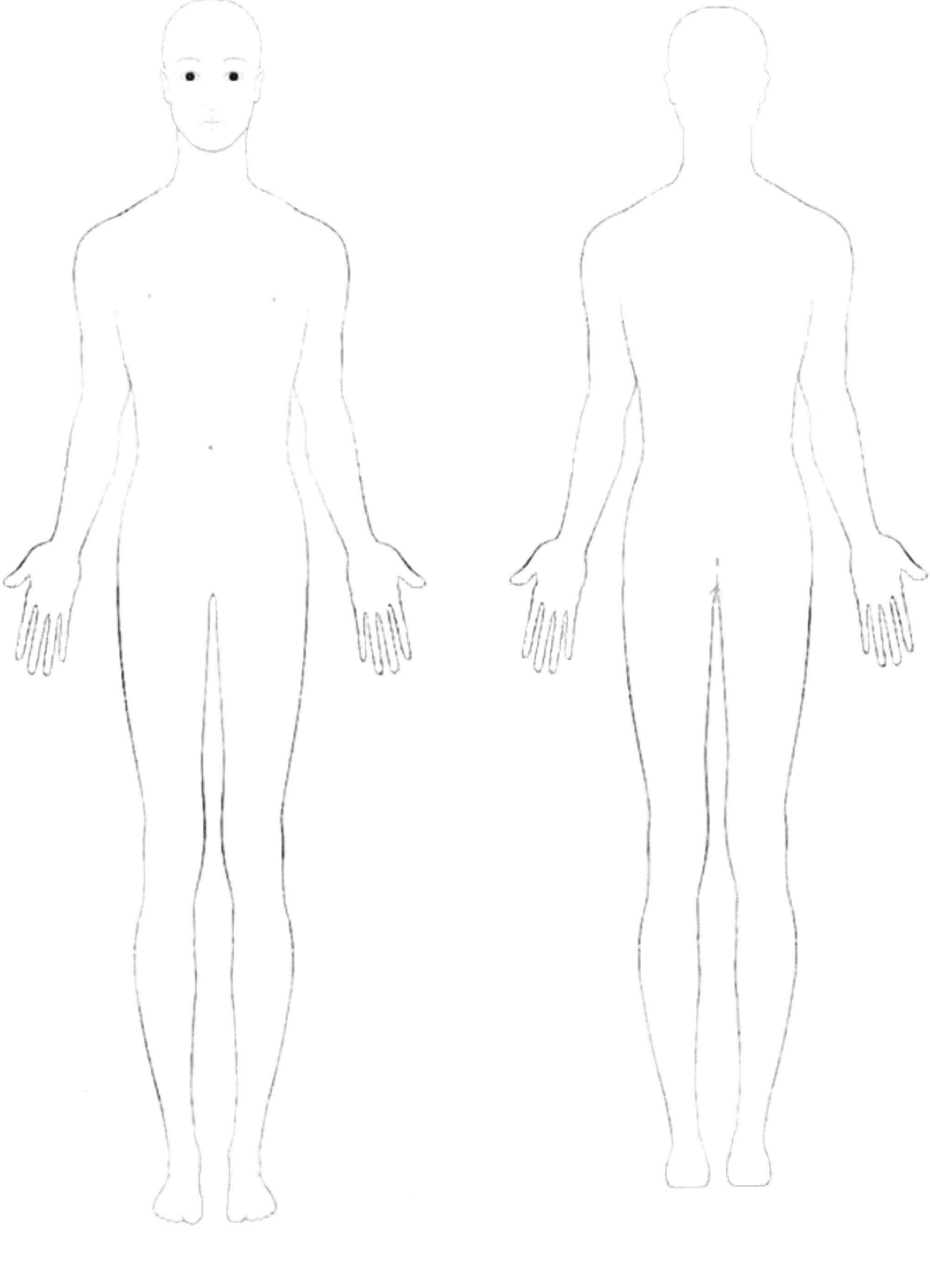

4. 분장 디자인 - 얼굴 Makeup Design - Face

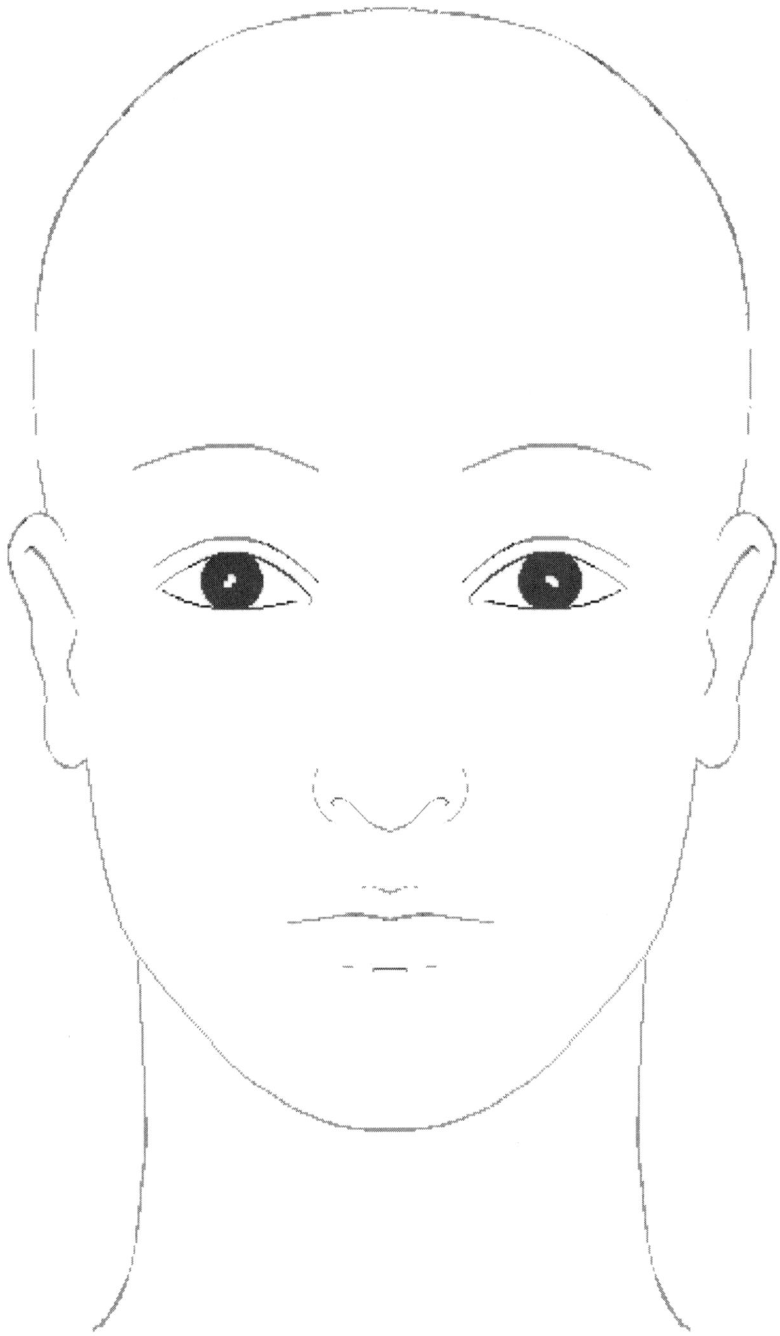

4. 분장 디자인 - 손, 발 Makeup Design - Hands, Feet

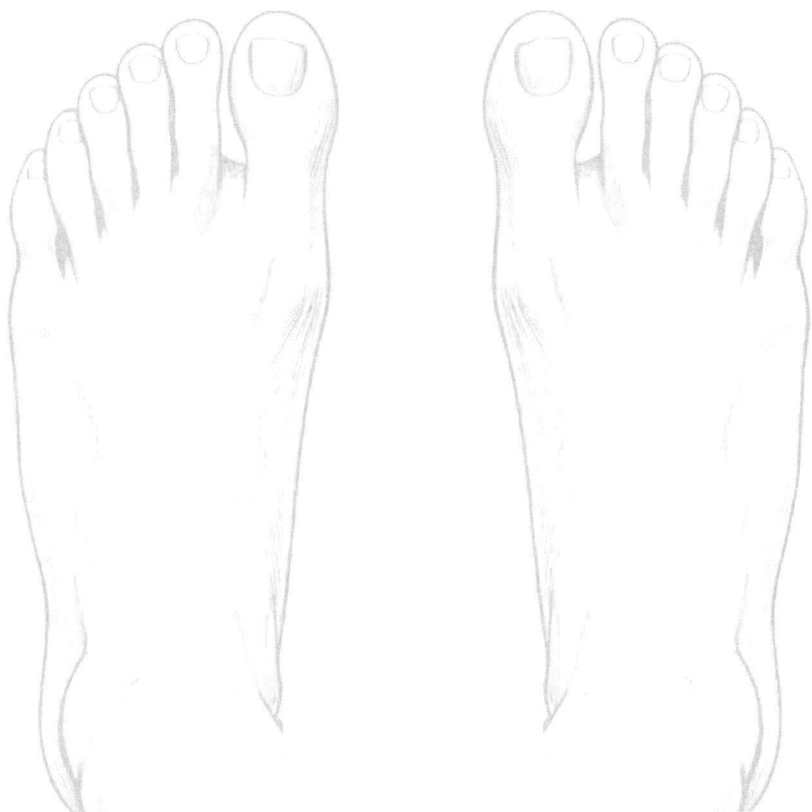

5. 동상 분장 완성 - 얼굴 Frostbite Makeup Completed - Face

완성 분장 이미지 삽입
Insert Completed Makeup Image.

5. 동상 분장 완성 - 손, 발 Frostbite Makeup Completed - Hands, Feet

완성 분장 이미지 삽입
Insert Completed Makeup Image.

6. 분장 계획 시행 Implementation of Makeup Plan

분장 계획 시행

Work Name 작품명		Actor 배우	

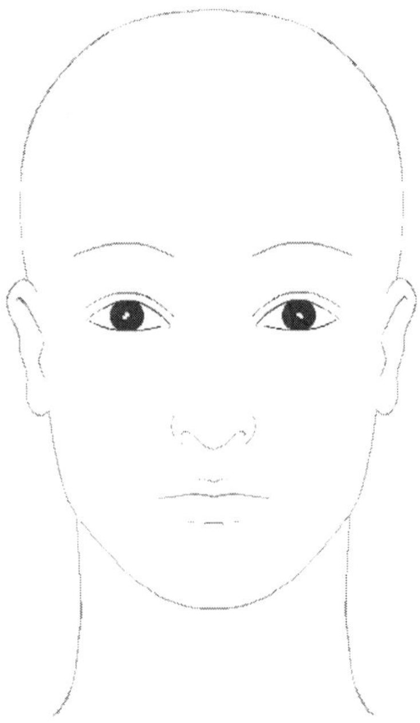

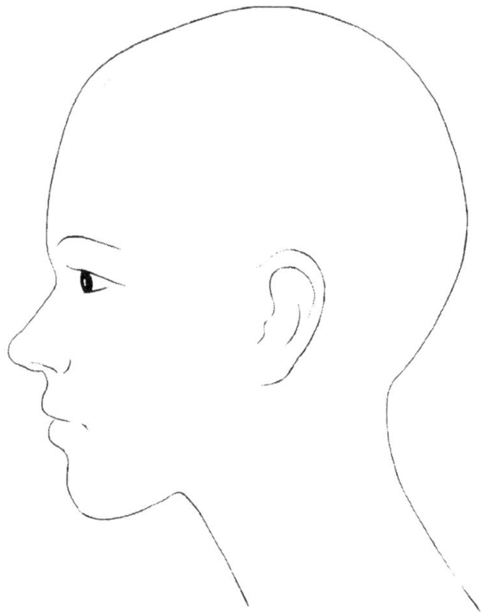

이마 Forehead	눈썹 Eyebrows	눈 Eye	코 Nose	뺨 Cheek
입 Mouth	화운데이션 Foundation	아이라인 Eye liners	가장 밝은 부분 Highlight	명암 Shading
파우더 Powder	목, 어깨 Neck, Shoulder	팔 Arm	복부, 배 Abdomen	등 Back
다리 Leg	손, 손가락 Hand, Finger	발, 발가락 Foot, Toe	헤어 Hair	그 외 메모사항 Other Notes

7. 등장인물 분석 Character Analysis

등장인물 분석	
작품명 Work Name	장르 Genre
배역명 Character Name	역할 Part
나이/성별 Age/Sex	성격 Personality
건강 Health	환경 Environment
유전학 Genetics	시대/국적 Times/Nationality
특징 Characteristic	

8. 분장의도 및 성격표현 Expression of Makeup Design and Character Portrayal

분장의도 및 성격표현		
	설명 Explain	만족도 Satisfaction
분장 의도 Makeup Purpose		
성격 표현 Character Expression		

Lesson 10

화상 분장
Burn Makeup

Lesson 10. 화상 분장 Burn Makeup

1. 자료조사 Research

• **화상 분장 재료 Burn Makeup Materials**

화상 분장 재료 이미지 삽입
Insert burn Makeup Materials Image.

2. 화상 Burn

- 화상의 단계는 일반적으로 네 가지로 나누며 각 단계는 화상의 깊이와 심각성에 따라 다르다.
  ① 화염 화상 : 화재 사고나 프로판, LPG 가스의 폭발로 인하여 화상
  ② 열탕 화상 : 뜨거운 물이나 식용유, 수증기 등에 의하여 화상
  ③ 전기 화상 : 감전으로 인해 발생하는 화상
  ④ 화학 화상 : 산, 알칼리 등 화학용품 접촉에 의한 화상
  ⑤ 접촉 화상 : 뜨거운 철판, 다리미, 전기장판 등에 의한 화상

1) 1도 화상 First-Degree Burn

- **특징**
- **증상**
- **치료**

화상 관련 자료 이미지 삽입
Insert Images Related to Burns.

2) 2도 화상 Second-Degree Burn

- **특징**
- **증상**
- **치료**

화상 관련 자료 이미지 삽입
Insert Images Related to Burns.

3) 3도 화상 Third-Degree Burn

- **특징**
- **증상**
- **치료**

화상 관련 자료 이미지 삽입
Insert Images Related to Burns.

4) 4도 화상 Fourth-Degree Burn

- **특징**
- **증상**
- **치료**

화상 관련 자료 이미지 삽입
Insert Images Related to Burns.

3. 분장 디자인 부위 Makeup Design Area

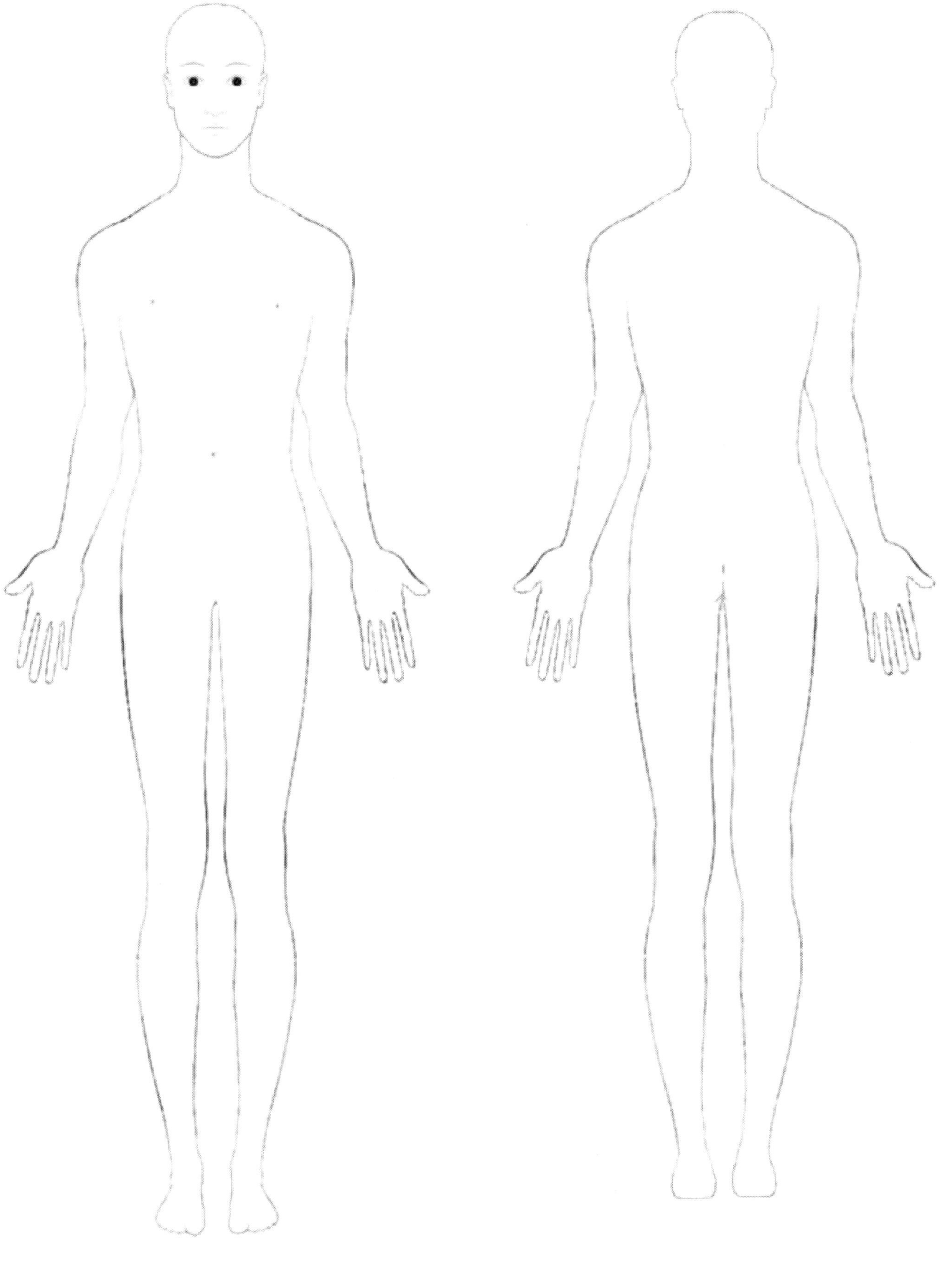

4. 분장 디자인 - 얼굴 Makeup Design - Face

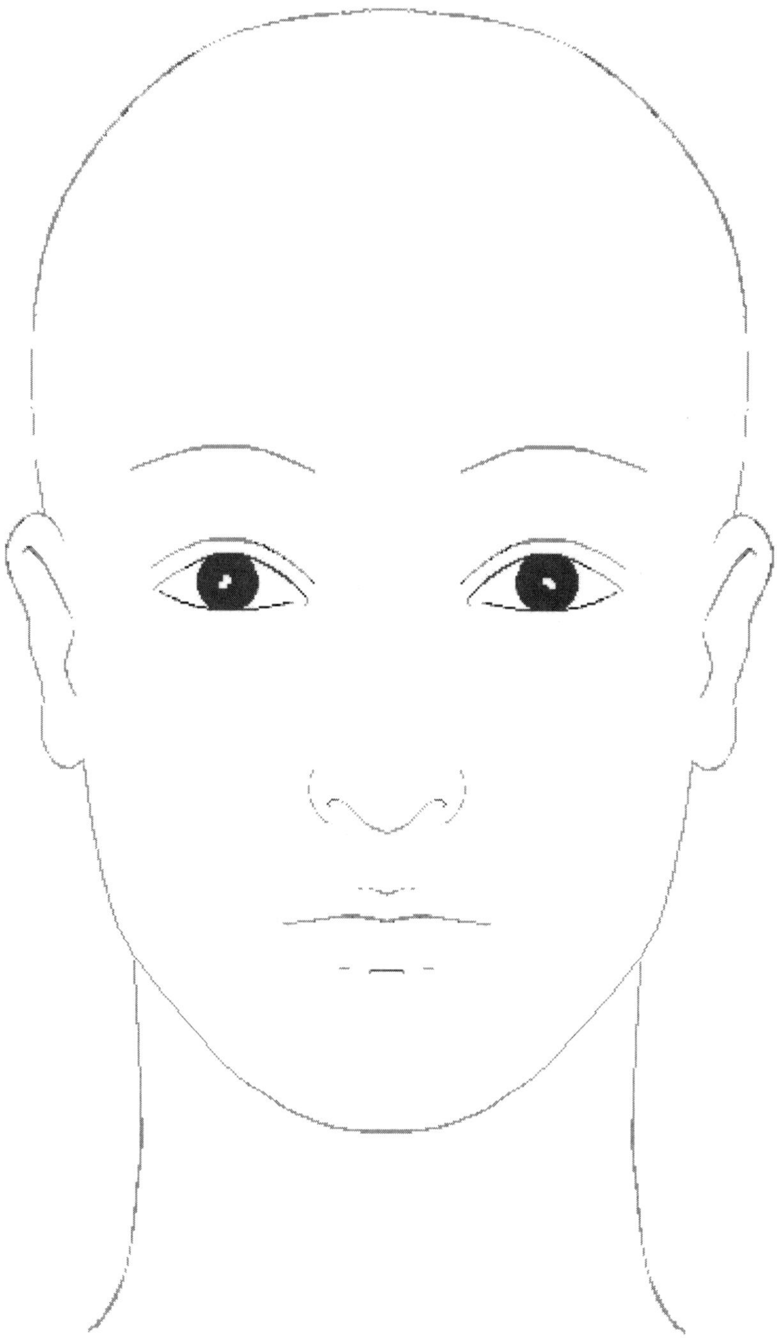

4. 분장 디자인 - 손, 발 Makeup Design - Hands, Feet

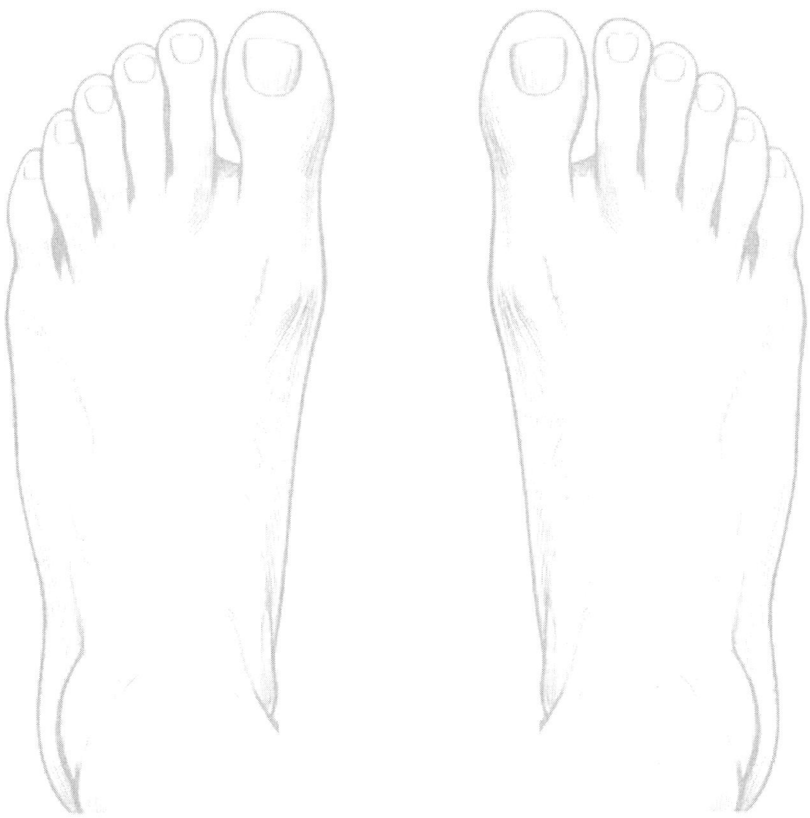

5. 화상 분장 완성 - 얼굴 Burn Makeup Completed - Face

완성 분장 이미지 삽입
Insert Completed Makeup Image.

5. 화상 분장 완성 - 신체 Burn Makeup Completed - Body

완성 분장 이미지 삽입
Insert Completed Makeup Image.

6. 분장 계획 시행 Implementation of Makeup Plan

분장 계획 시행			
Work Name 작품명		Actor 배우	

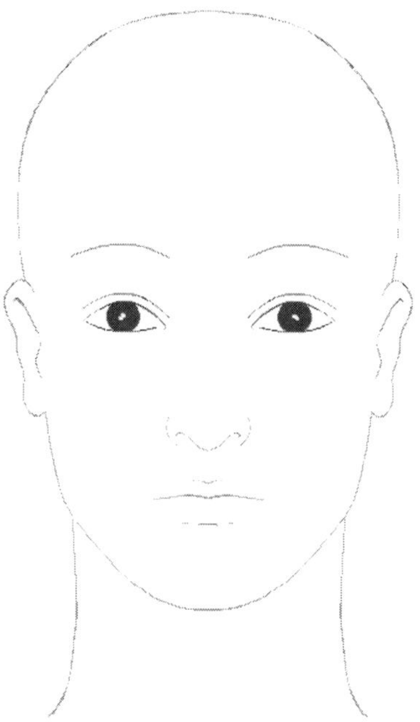

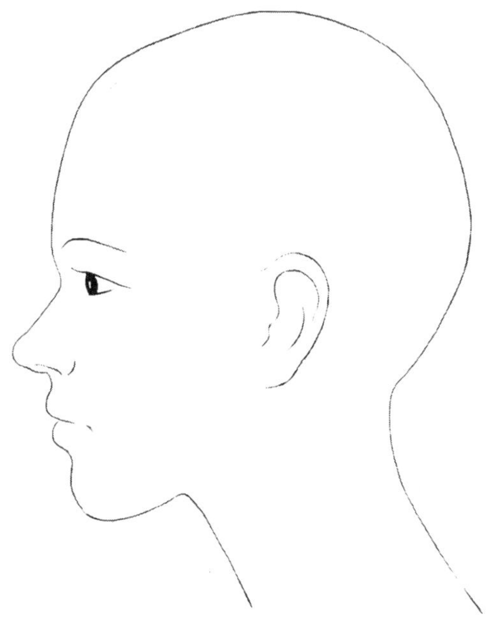

이마 Forehead	눈썹 Eyebrows	눈 Eye	코 Nose	뺨 Cheek
입 Mouth	화운데이션 Foundation	아이라인 Eye liners	가장 밝은 부분 Highlight	명암 Shading
파우더 Powder	목, 어깨 Neck, Shoulder	팔 Arm	복부, 배 Abdomen	등 Back
다리 Leg	손, 손가락 Hand, Finger	발, 발가락 Foot, Toe	헤어 Hair	그 외 메모사항 Other Notes

7. 등장인물 분석 Character Analysis

<table>
<tr><th colspan="2">등장인물 분석</th></tr>
<tr><td>작품명 Work Name</td><td>장르 Genre</td></tr>
<tr><td>배역명 Character Name</td><td>역할 Part</td></tr>
<tr><td>나이/성별 Age/Sex</td><td>성격 Personality</td></tr>
<tr><td>건강 Health</td><td>환경 Environment</td></tr>
<tr><td>유전학 Genetics</td><td>시대/국적 Times/Nationality</td></tr>
<tr><td colspan="2">특징 Characteristic</td></tr>
</table>

8. 분장의도 및 성격표현 Expression of Makeup Design and Character Portrayal

<table>
<tr><th colspan="3">분장의도 및 성격표현</th></tr>
<tr><td></td><td>설명 Explain</td><td>만족도 Satisfaction</td></tr>
<tr><td>분장 의도
Makeup
Purpose</td><td></td><td></td></tr>
<tr><td>성격 표현
Character
Expression</td><td></td><td></td></tr>
</table>

Lesson 11

질병 분장
Disease Makeup

Lesson 11. 질병 분장 Disease Makeup

1. 자료조사

1) 질병 분장 완성

- 특정 질병이나 상태를 시각적으로 표현하는 기술로, 영화, TV드라마, 연극, 사진 촬영 등에서 사용된다.
- 피부 변화, 상처, 색 변화 등을 포함해 실제 질병의 증상 및 특성을 재현하는 것이다.
- 실제 환자의 증상, 치료 과정 등 특정 질병에 대한 이해가 필요하다.
- 윤리적 고려 : 질병을 다룰 때는 민감함이 필요하며, 환자에게 불쾌감을 주지 않도록 주의해야 한다.

2) 질병 자료조사

질병관련 자료 이미지 삽입
Insert Images Related to the disease.

2. 질병 분장 디자인

1안 Option1

2안 Option2

3. 질병 분장 완성 Disease Makeup Completed

완성 분장 정면 이미지 삽입
Insert Completed Makeup Front View Image.

완성 분장 좌측면 이미지 삽입
Insert Completed Makeup left Side View Image.

완성 분장 우측면 이미지 삽입
Insert Completed Makeup Right Side View Image.

4. 분장 디자인 부위 Makeup Design Area

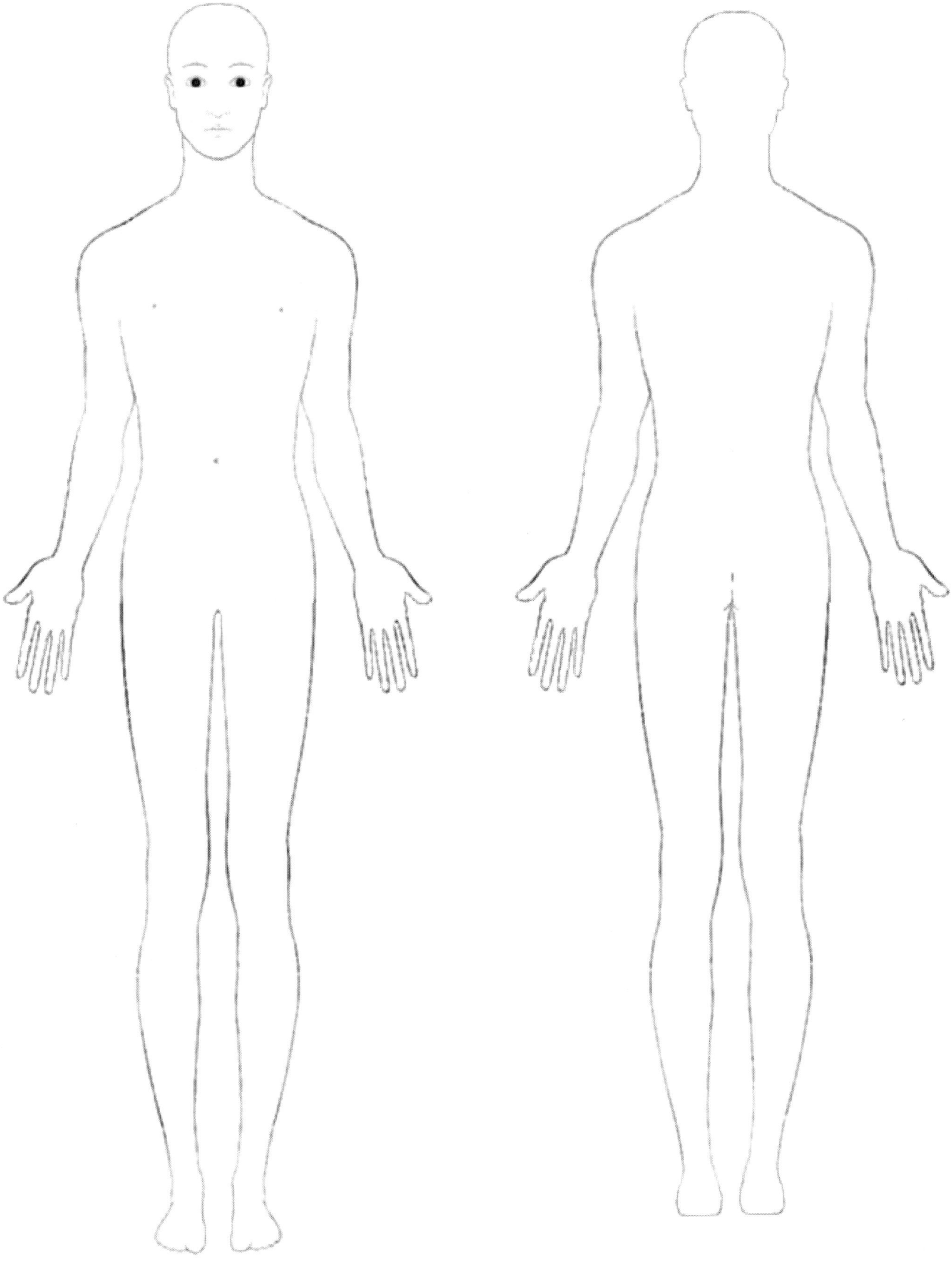

5. 분장 계획 시행 Implementation of Makeup Plan

분장 계획 시행			
Work Name 작품명		Actor 배우	

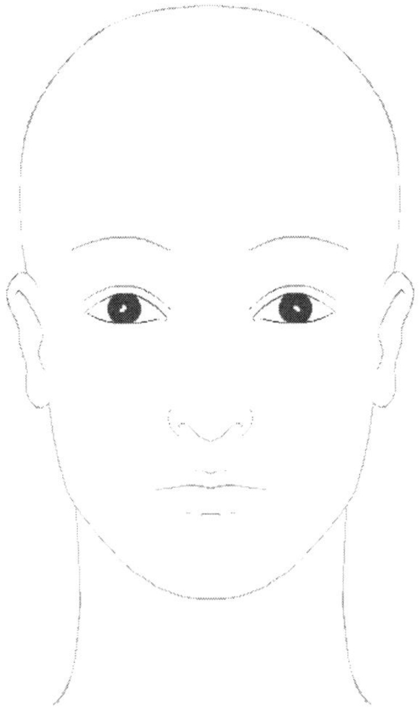

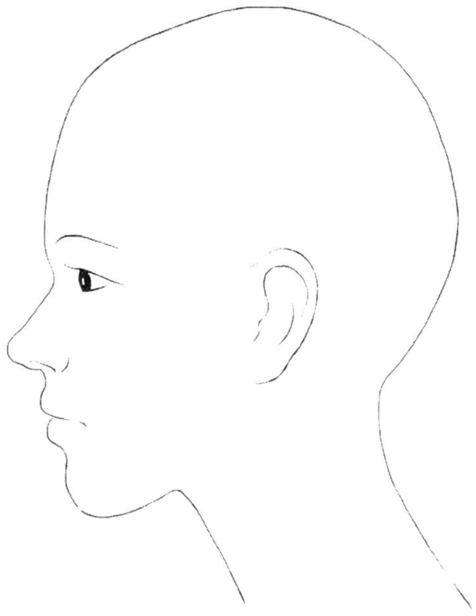

이마 Forehead	눈썹 Eyebrows	눈 Eye	코 Nose	뺨 Cheek
입 Mouth	화운데이션 Foundation	아이라인 Eye liners	가장 밝은 부분 Highlight	명암 Shading
파우더 Powder	목, 어깨 Neck, Shoulder	팔 Arm	복부, 배 Abdomen	등 Back
다리 Leg	손, 손가락 Hand, Finger	발, 발가락 Foot, Toe	헤어 Hair	그 외 메모사항 Other Notes

6. 등장인물 분석 Character Analysis

<table>
<tr><th colspan="2">등장인물 분석</th></tr>
<tr><td>작품명 Work Name</td><td>장르 Genre</td></tr>
<tr><td>배역명 Character Name</td><td>역할 Part</td></tr>
<tr><td>나이/성별 Age/Sex</td><td>성격 Personality</td></tr>
<tr><td>건강 Health</td><td>환경 Environment</td></tr>
<tr><td>유전학 Genetics</td><td>시대/국적 Times/Nationality</td></tr>
<tr><td colspan="2">특징 Characteristic</td></tr>
</table>

7. 분장의도 및 성격표현 Expression of Makeup Design and Character Portrayal

<table>
<tr><th colspan="3">분장의도 및 성격표현</th></tr>
<tr><td></td><td>설명 Explain</td><td>만족도 Satisfaction</td></tr>
<tr><td>분장 의도
Makeup
Purpose</td><td></td><td></td></tr>
<tr><td>성격 표현
Character
Expression</td><td></td><td></td></tr>
</table>

Lesson 12

캐릭터 분장 디자인 및 연출
Character Makeup Design and Direction

Lesson 12. 캐릭터 분장 디자인 및 연출 Character Makeup Design and Direction

1. 분장 디자인 영상 촬영 계획 Makeup Design Video Shooting Plan

팀 명 Team Name		작품명 Title of the Work	
팀 구성 Team Members	구성원 Members	등장인물 분석 Character Analysis	
	□ 팀장 Team Leader :		
	□ 조원 Team Members :		
	□ 조원 Team Members :		
	□ 조원 Team Members :		
	□ 조원 Team Members :		
	□ 조원 Team Members :		
	□ 조원 Team Members :		
촬영장소 Filming Location			
촬영일자 Filming Date			
줄거리 Synopsis			

2. 시나리오 Scenario or Script

3. 분장 연출 계획 Makeup Production Plan

팀 명 Team Name	

분장	□ 스트레이트 분장	□ 노인분장	□ 타박상 분장
	□ 봉합상처 분장	□ 찰과상 분장	□ 열상 분장
	□ 창상 분장	□ 총상 분장	□ 자상 분장
	□ 대머리 분장	□ 수염 분장	□ 동상 분장
	□ 화상 분장	□ 질병 분장	□

No	분장사 명 Makeup Artist Name	담당 분장 Assigned Makeup	분장 상황 Makeup Situation
1			
2			
3			
4			
5			
6			
7			
8			

4. 분장 디자인 시안 - 얼굴1 Face Makeup Design Draft1

등장인물 분석
분장명
배역명 Character Name
나이/성별 Age/Sex
역할 Part
성격 Personality
유전학 Genetics
건강 Health
환경 Environment
시대/국적 Times/Nationality
분장 의도 Makeup Purpose

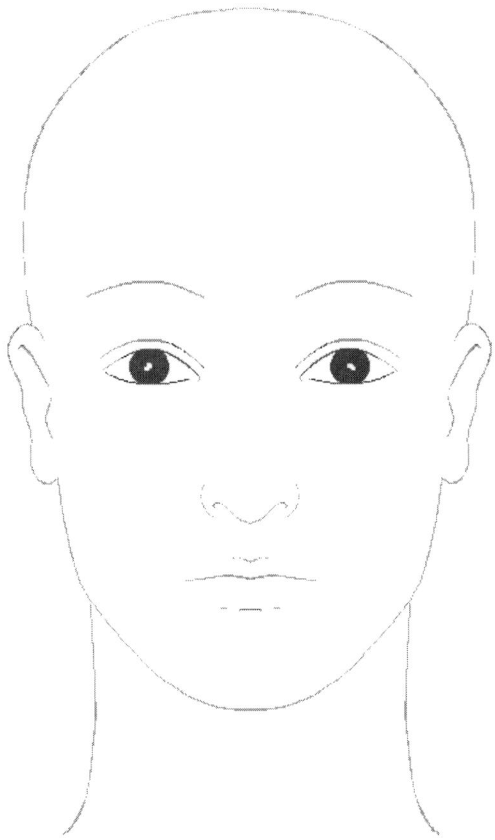

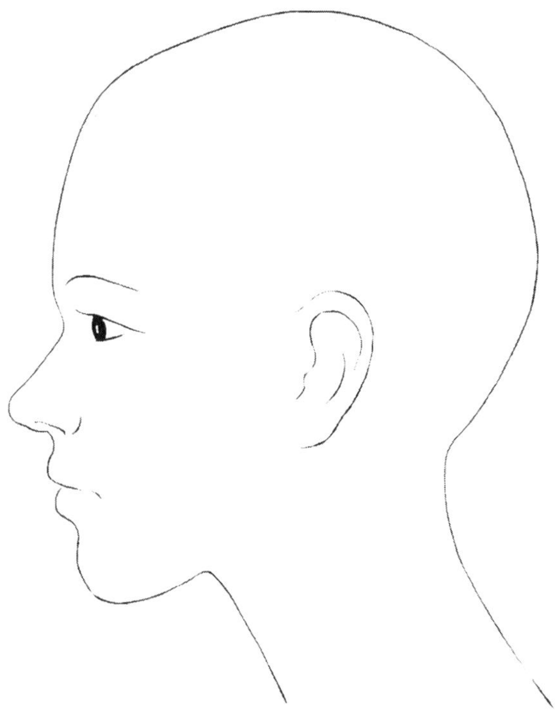

4. 분장 디자인 시안 - 전신1 Full Body Makeup Design Draft1

등장인물 분석

분장명

배역명 Character Name

나이/성별 Age/Sex

역할 Part

성격 Personality

유전학 Genetics

건강 Health

환경 Environment

시대/국적 Times/Nationality

분장 의도 Makeup Purpose

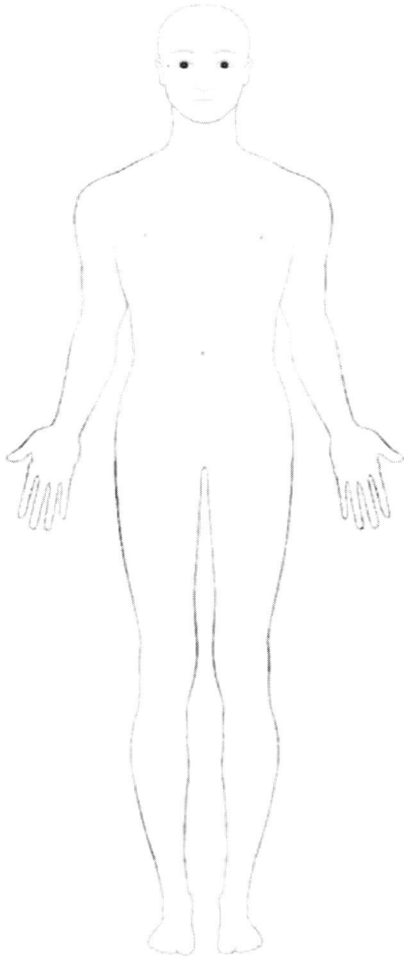

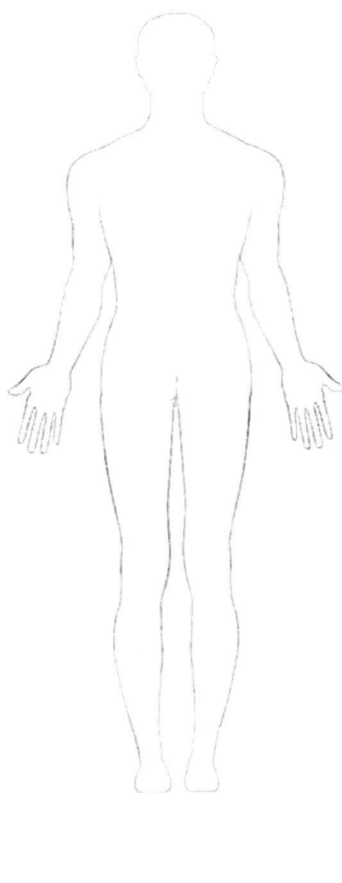

5. 분장 완성1 Makeup Completed1

완성 분장 정면 이미지 삽입
Insert Completed Makeup Front View Image.

완성 분장 좌측면 이미지 삽입 Insert Completed Makeup left Side View Image.	완성 분장 우측면 이미지 삽입 Insert Completed Makeup Right Side View Image.

6. 분장 계획 시행1 Implementation of Makeup Plan1

적용분장 Applied Makeup	분장사 Makeup Artist	모델명 Model Name

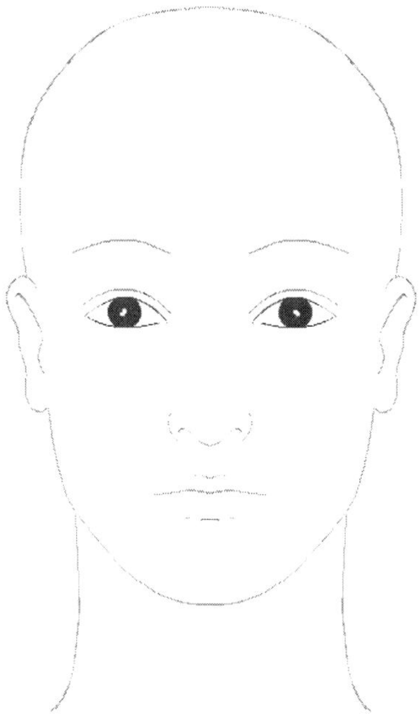

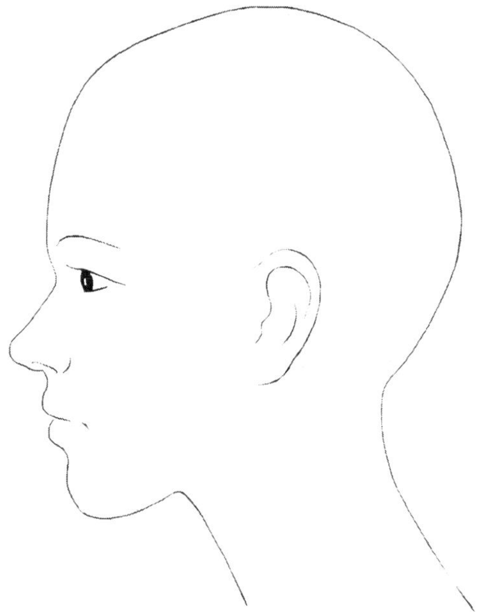

이마 Forehead	눈썹 Eyebrows	눈 Eye	코 Nose	뺨 Cheek
입 Mouth	화운데이션 Foundation	아이라인 Eye liners	가장 밝은 부분 Highlight	명암 Shading
파우더 Powder	목, 어깨 Neck, Shoulder	팔 Arm	복부, 배 Abdomen	등 Back
다리 Leg	손, 손가락 Hand, Finger	발, 발가락 Foot, Toe	헤어 Hair	그 외 메모사항 Other Notes

7. 등장인물 분석1 Character Analysis1

등장인물 분석	
작품명 Work Name	장르 Genre
배역명 Character Name	역할 Part
나이/성별 Age/Sex	성격 Personality
건강 Health	환경 Environment
유전학 Genetics	시대/국적 Times/Nationality
특징 Characteristic	

8. 분장의도 및 성격표현1 Expression of Makeup Design and Character Portrayal1

분장의도 및 성격표현		
	설명 Explain	만족도 Satisfaction
분장 의도 Makeup Purpose		
성격 표현 Character Expression		

4. 분장 디자인 시안 - 얼굴2 Face Makeup Design Draft2

등장인물 분석

분장명

배역명 Character Name

나이/성별 Age/Sex

역할 Part

성격 Personality

유전학 Genetics

건강 Health

환경 Environment

시대/국적 Times/Nationality

분장 의도 Makeup Purpose

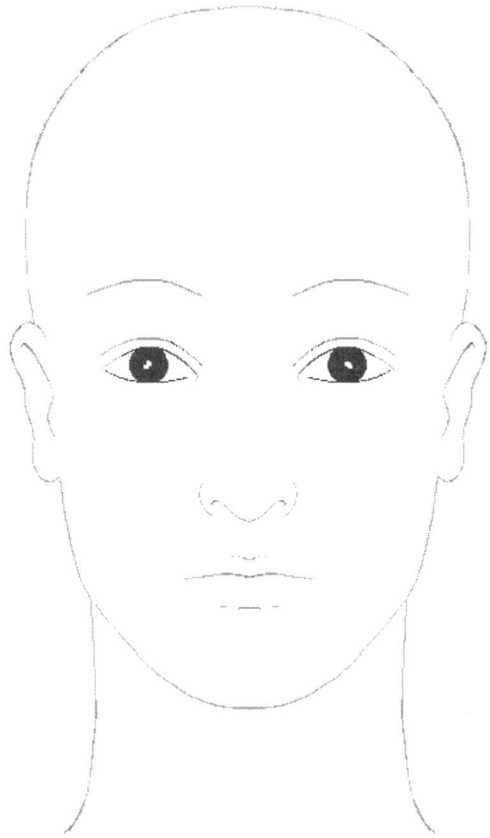

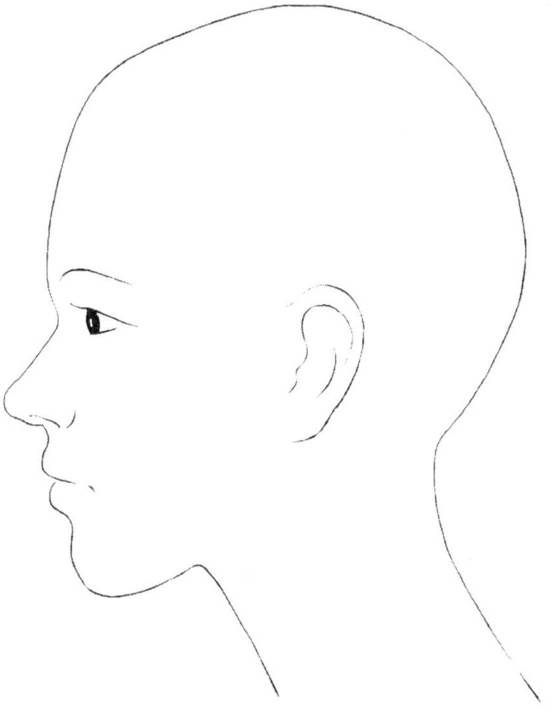

4. 분장 디자인 시안 - 전신2 Full Body Makeup Design Draft2

등장인물 분석

분장명

배역명 Character Name

나이/성별 Age/Sex

역할 Part

성격 Personality

유전학 Genetics

건강 Health

환경 Environment

시대/국적 Times/Nationality

분장 의도 Makeup Purpose

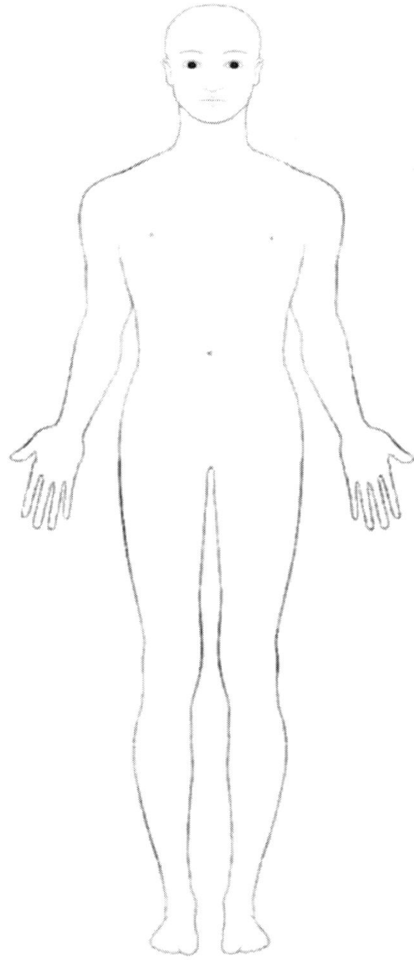

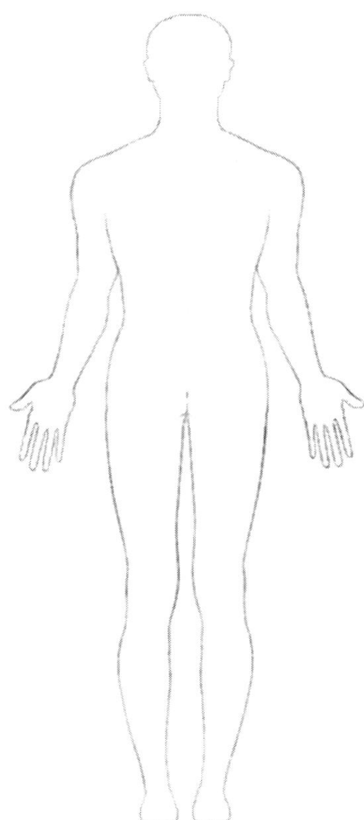

5. 분장 완성2 Makeup Completed2

완성 분장 정면 이미지 삽입
Insert Completed Makeup Front View Image.

완성 분장 좌측면 이미지 삽입 Insert Completed Makeup left Side View Image.	완성 분장 우측면 이미지 삽입 Insert Completed Makeup Right Side View Image.

6. 분장 계획 시행2 Implementation of Makeup Plan2

적용분장 Applied Makeup	분장사 Makeup Artist	모델명 Model Name

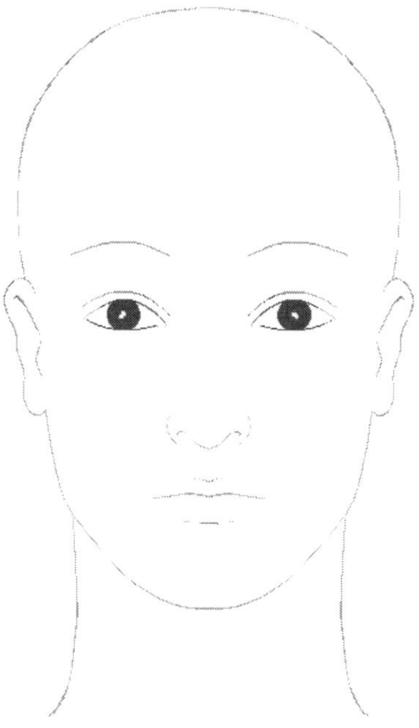

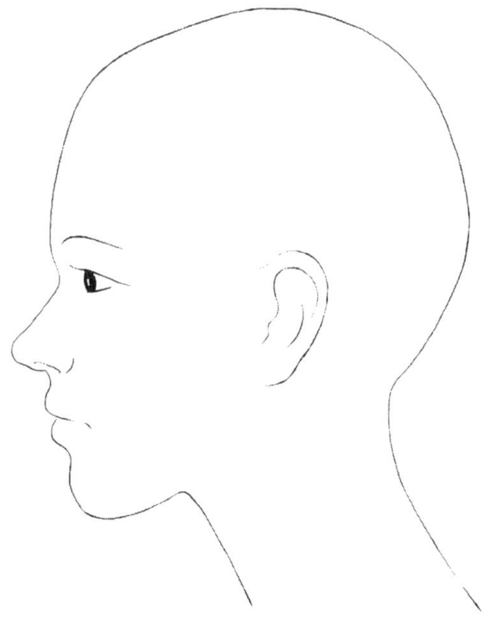

이마 Forehead	눈썹 Eyebrows	눈 Eye	코 Nose	뺨 Cheek
입 Mouth	화운데이션 Foundation	아이라인 Eye liners	가장 밝은 부분 Highlight	명암 Shading
파우더 Powder	목, 어깨 Neck, Shoulder	팔 Arm	복부, 배 Abdomen	등 Back
다리 Leg	손, 손가락 Hand, Finger	발, 발가락 Foot, Toe	헤어 Hair	그 외 메모사항 Other Notes

7. 등장인물 분석2 Character Analysis2

등장인물 분석	
작품명 Work Name	장르 Genre
배역명 Character Name	역할 Part
나이/성별 Age/Sex	성격 Personality
건강 Health	환경 Environment
유전학 Genetics	시대/국적 Times/Nationality
특징 Characteristic	

8. 분장의도 및 성격표현2 Expression of Makeup Design and Character Portrayal2

분장의도 및 성격표현		
	설명 Explain	만족도 Satisfaction
분장 의도 Makeup Purpose		
성격 표현 Character Expression		

4. 분장 디자인 시안 - 얼굴3 Face Makeup Design Draft3

등장인물 분석
분장명
배역명 Character Name
나이/성별 Age/Sex
역할 Part
성격 Personality
유전학 Genetics
건강 Health
환경 Environment
시대/국적 Times/Nationality
분장 의도 Makeup Purpose

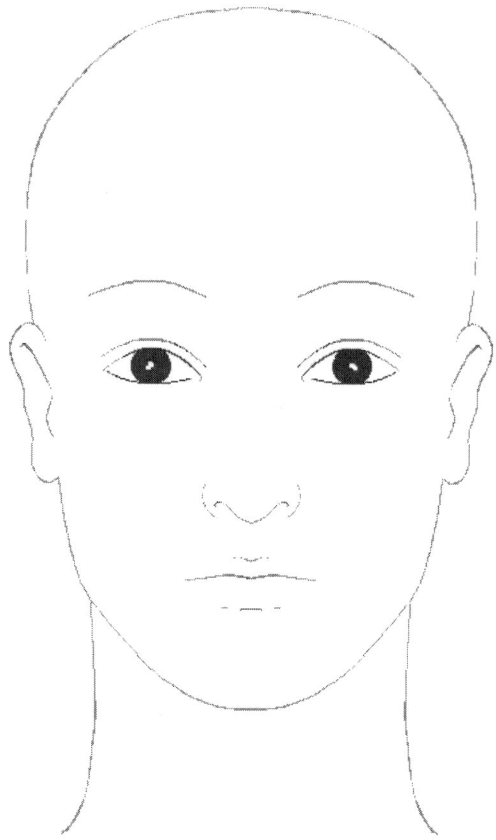

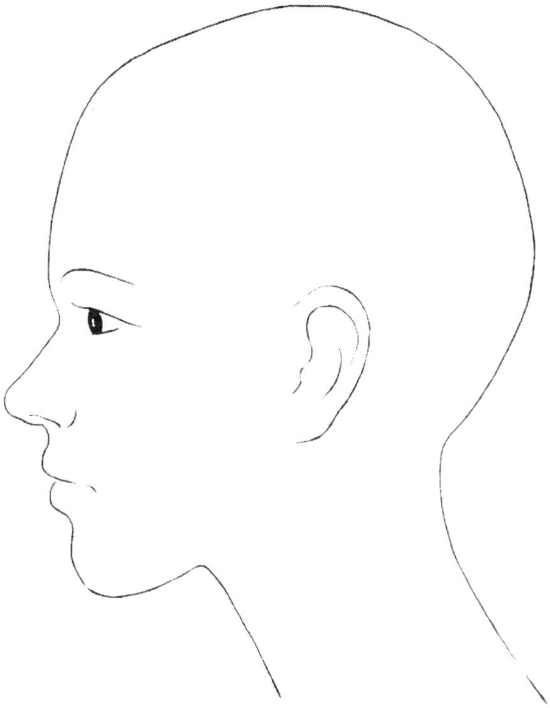

4. 분장 디자인 시안 - 전신3 Full Body Makeup Design Draft3

등장인물 분석
분장명
배역명 Character Name
나이/성별 Age/Sex
역할 Part
성격 Personality
유전학 Genetics
건강 Health
환경 Environment
시대/국적 Times/Nationality
분장 의도 Makeup Purpose

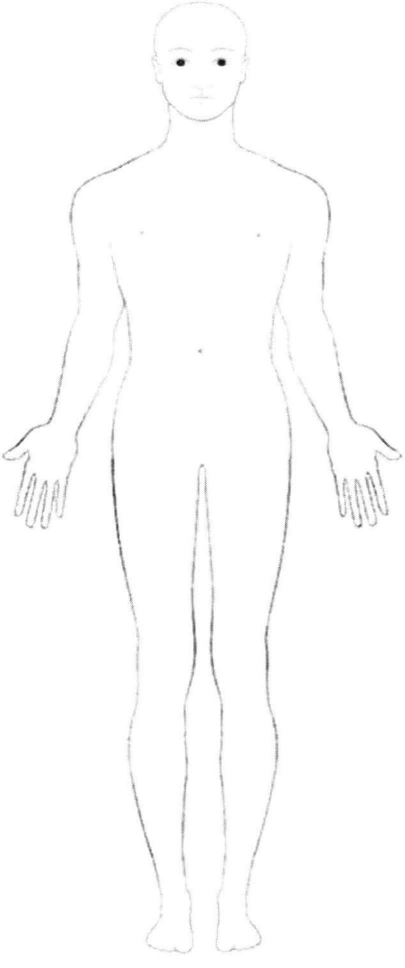

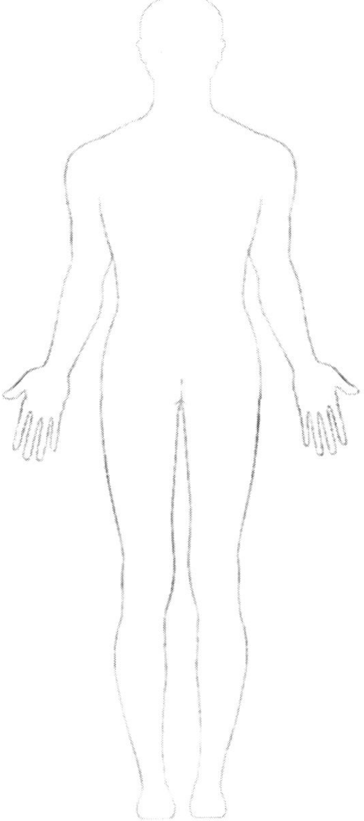

5. 분장 완성3 Makeup Completed3

완성 분장 정면 이미지 삽입
Insert Completed Makeup Front View Image.

완성 분장 좌측면 이미지 삽입 Insert Completed Makeup left Side View Image.	완성 분장 우측면 이미지 삽입 Insert Completed Makeup Right Side View Image.

6. 분장 계획 시행3 Implementation of Makeup Plan3

적용분장 Applied Makeup	분장사 Makeup Artist	모델명 Model Name

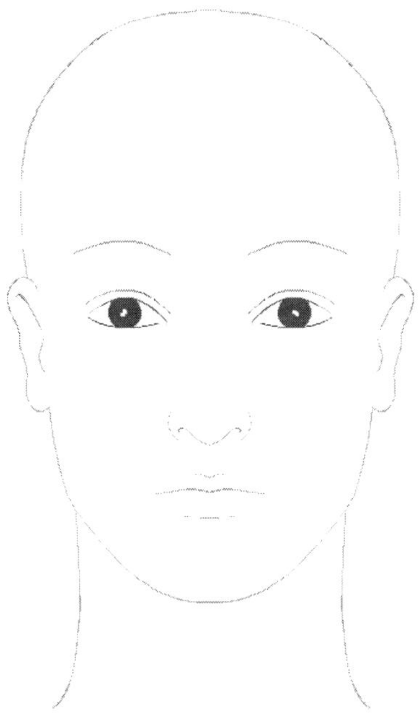

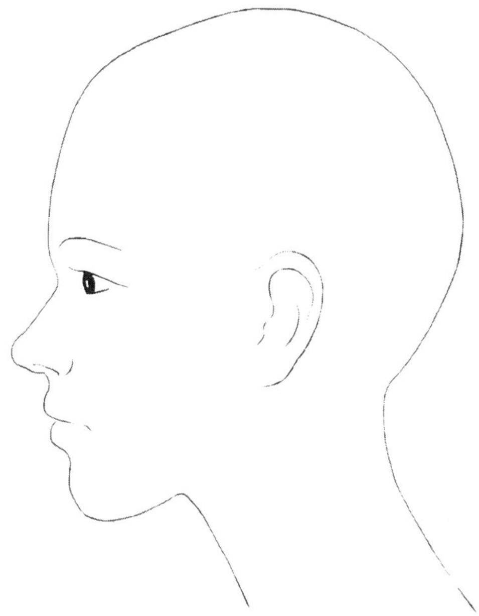

이마 Forehead	눈썹 Eyebrows	눈 Eye	코 Nose	뺨 Cheek
입 Mouth	화운데이션 Foundation	아이라인 Eye liners	가장 밝은 부분 Highlight	명암 Shading
파우더 Powder	목, 어깨 Neck, Shoulder	팔 Arm	복부, 배 Abdomen	등 Back
다리 Leg	손, 손가락 Hand, Finger	발, 발가락 Foot, Toe	헤어 Hair	그 외 메모사항 Other Notes

7. 등장인물 분석3 Character Analysis3

등장인물 분석	
작품명 Work Name	장르 Genre
배역명 Character Name	역할 Part
나이/성별 Age/Sex	성격 Personality
건강 Health	환경 Environment
유전학 Genetics	시대/국적 Times/Nationality
특징 Characteristic	

8. 분장의도 및 성격표현3 Expression of Makeup Design and Character Portrayal3

분장의도 및 성격표현		
	설명 Explain	만족도 Satisfaction
분장 의도 Makeup Purpose		
성격 표현 Character Expression		

참고문헌

1. Bobbi Brown(2011), Bobbi Brown Makeup Manual: For Everyone from Beginner to Pro, Grand Central Life & Style; Reprint edition.
2. Faber Birren(2013), Color Psychology and Color Therapy
3. Leatrice Eiseman(2000), Pantone Guide to Communicating with Color
4. Richard Corson(2001), Stage Makeup, Pearson Custom Publishing.
5. Todd Debreceni(2018), Special Makeup Effects for Stage and Screen, Routledge; 3rd edition.
6. 교육부(2019), 미디어 메이크업(LM12010348_18v4), 한국직업능력개발원.
7. 국립극장(2024), 무대조명, 교보문고.
8. 윤예령·강지연·이선주·임희경·최정아(2022), 미디어메이크업, 메디시안.
9. 임희경·HUO TAO(2021), Stage Make-up, 청구문화사.
10. 네이버사전 https://dict.naver.com/
11. 브리태니커 백과사전 https://www.britannica.com/
12. Wikipedia, 동상, https://ko.wikipedia.org/wiki/%EB%8F%99%EC%83%81
13. 서울아산병원, 동상, https://www.amc.seoul.kr/asan/search/search.do?kwd=%EB%8F%99%EC%83%81
14. Wikipedia, 화상, https://ko.wikipedia.org/wiki/%ED%99%94%EC%83%81
15. 서울아산병원, 화상, https://www.amc.seoul.kr/asan/healthinfo/disease/diseaseDetail.do?contentId=32321